OPERATIONAL EFFICIENCY IN FORESTRY
VOL. 2: PRACTICE

FORESTRY SCIENCES

Baas P, ed: New Perspectives in Wood Anatomy. 1982. ISBN 90-247-2526-7
Prins CFL, ed: Production, Marketing and Use of Finger-Jointed Sawnwood. 1982. ISBN 90-247-2569-0
Oldeman RAA, et al., eds: Tropical Hardwood Utilization: Practice and Prospects. 1982. ISBN 90-247-2581-X
Den Ouden P and Boom BK: Manual of Cultivated Conifers: Hardy in Cold and Warm-Temperate Zone. 1982. ISBN 90-247-2148-2
Bonga JM and Durzan DJ, eds: Tissue Culture in Forestry. 1982. ISBN 90-247-2660-3
Satoo T and Magwick HAI: Forest Biomass. 1982. ISBN 90-247-2710-3
Van Nao T, ed: Forest Fire Prevention and Control. 1982. ISBN 90-247-3050-3
Douglas J: A Re-appraisal of Forestry Development in Developing Countries. 1983. ISBN 90-247-2830-4
Gordon JC and Wheeler CT, eds: Biological Nitrogen Fixation in Forest Ecosystems: Foundations and Applications. 1983. ISBN 90-247-2849-5
Németh MV: The Virus-Mycoplasma and Rikettsia Disease of Fruit Trees. ISBN 90-247-2868-1
Duryea ML and Landis TD, eds: Forest Nursery Manual: Production of Bareroot Seedlings. 1984. ISBN 90-247-2913-0
Hummel FC, ed: Forest Policy: A Contribution to Resource Development. 1984. ISBN 90-247-2883-5
Manion PD, ed: Scleroderris Canker of Conifers. 1984. ISBN 90-247-2912-2
Duryea ML and Brown GN, eds: Seedling Physiology and Reforestation Success. 1984. ISBN 90-247-2949-1
Staaf KAG and Wiksten NA: Tree Harvesting Techniques. 1984. ISBN 90-247-2994-7
Boyd JD: Biophysical Control of Microfibril Orientation in Plant Cell Walls. 1985. ISBN 90-247-3101-1
Findlay WPK, ed: Preservation of Timber in the Tropics. 1985. ISBN 90-247-3112-7
Samset I: Winch and Cable Systems. 1985. ISBN 90-247-3205-0
Leary RA: Interaction Theory in Forest Ecology and Management. 1985. ISBN 90-247-3220-4
Gessel SP: Forest Site and Productivity. 1986. ISBN 90-247-3284-0
Hennessey TC, Dougherty PM, Kossuth SV and Johnson JD, eds: Stress Physiology and Forest Productivity. 1986. ISBN 90-247-3359-6
Shepherd KR: Plantation Silviculture. 1986. ISBN 90-247-3379-0
Sohlberg S and Sokolov VE, eds: Practical Application of Remote Sensing in Forestry. 1986. ISBN 90-247-3392-8
Bonga JM and Durzan DJ, eds: Cell and Tissue Culture in Forestry. Volume 1: General Principles and Biotechnology. 1987. ISBN 90-247-3430-4
Bonga JM and Durzan DJ, eds: Cell and Tissue Culture in Forestry. Volume 2: Specific Principles and Methods: Growth and Developments. 1987. ISBN 90-247-3431-2
Bonga JM and Durzan DJ, eds: Cell and Tissue Culture in Forestry. Volume 3: Case Histories: Gymnosperms, Angiosperms and Palms. 1987. ISBN 90-247-3432-0
(Set Vols. 1–3: ISBN 90-247-3433-9)
Richards EG, ed: Forestry and the Forest Industries: Past and Future. 1987. ISBN 90-247-3592-0
Kossuth SV and Ross SD, eds: Hormonal Control of Tree Growth. 1987. ISBN 90-247-3621-8
Sundberg U and Silversides CR, eds: Operational Efficiency in Forestry. Volume 1: Analysis. 1988. ISBN 90-247-3683-8
Silversides CR and Sundberg U: Operational Efficiency in Forestry. Volume 2: Practice. 1989. ISBN 0-7923-0063-7

Operational Efficiency in Forestry

Vol. 2: Practice

by

C.R. SILVERSIDES
Prescott, Ontario, Canada

and

U. SUNDBERG
Swedish University of Agricultural Sciences, Garpenberg, Sweden

SPRINGER-SCIENCE+BUSINESS MEDIA, B.V.

Library of Congress Cataloging in Publication Data
(Revised for vol. 2)

Operational efficiency in forestry.

(Forestry sciences ; 29: v. 1; 32: v. 2)
Editors' names in reverse order in v. 2.
Bibliography: p.
Contents: v. 1. Analysis -- v. 2. Practice.
1. Forestry efficiency. 2. Forests and forestry--
Labor productivity. 3. Forestry engineering.
I. Sundberg, U. (Ulf) II. Silversides, C. R.
III. Series: Forestry sciences ; v. 29: v. 1; 32: v. 2, etc.
SD387.E33064 1988 338.1'6 88-2940

ISBN 978-90-481-4037-4 ISBN 978-94-017-0506-6 (eBook)
DOI 10.1007/978-94-017-0506-6

2nd printing 1996

PREFACE

It is our conviction that professional skill in forestry will develop more readily and more efficiently if forestry students are presented with a clear understanding of the impact of the physical factors that both enhance and inhibit forestry activities.

Part I is analytic, written as a basic text for undergraduates in courses such as logging, transport, forest engineering and even forest management. It deals with the fundamentals of technology in forestry as determined by the physical environment. The analytic approach serves two purposes, to bring about a clear understanding of the real world of the forest and to develop tools through which efficiency and productivity can be explored, understood and improved. The principal author of this volume was Prof. Ulf Sundberg.

Part II discusses in some detail a wide variety of practical problems encountered by foresters. It describes harvesting systems and the principles of management and control of forest operations. The influence of the forest on operations is described at length, the terrain, topography, forest soils as well as the engineering characteristics of trees and forest stands. It also considers the impact of operations on the forest. The principal author of this volume was Dr. Ross Silversides. Chapters 11, 12 and 13 were written by Prof. Sundberg.

With these two volumes our aim has been to produce a text with the essential technological knowledge that we consider every forester should have. The professional profile even of those specializing in biological and environmental subjects, should include an understanding of the technology of forest operations. No detailed descriptions of equipment are given as these are readily available elsewhere. In this period of mechanization of harvesting operations obsolescence is a dominant factor as new concepts for systems and system components appear and disappear at a rate unheard of pre-World War II.

Both authors wish to acknowledge the tremendous assistance afforded us by Gunvor Harrysson, Agneta Krohn, Kerstin Tordmar, and the illustrators, Britt Lindblad-Sundberg and Sigge Falk. Professor Per Olov Nilsson, Professor of Forest Energy Systems at Garpenberg, is sincerely thanked for permitting the use of the facilities of his department. We are indebted to the Central Fund for Employees in Forestry and Forest Industries for a grant which greatly assisted in the preparation of these volumes. Finally we would like to acknowledge the devoted support given by our wives, without which the volumes would not have materialized.

Prescott, Ontario and Krylbo, Sweden, 1988

C.R. Silversides and Ulf Sundberg

CONTENTS

FIGURES

Volume 1

Volume 2

Page

TABLES

Volume 1

Volume 2

Page

1. INTRODUCTION

It was pointed out in Part I that the basic economic problem in forest management is to achieve the most efficient use of productive resources. It is the need to manipulate and use the forest resource with a minimum of human activities that is the concept of operational efficiency with which this book is concerned.

Productivity is output and as growth in a forest it may be expressed as m^3/ha/yr or on harvesting and transport operations as m^3/h or tkm/h. The capacity of an operation is the quantity of variable items which can be combined with a fixed item to result in the optimum combination. If a tree harvester is operated within a forest stand its capacity will be affected by variation in tree size, in machine feed speed, etc. Productivity depends upon capacity. Another measure of quality is efficiency which is an indication of the effectiveness of an operation. Efficiency is always expressed as a rate, such as production in m^3/h or engine efficiency in kW/h/l fuel.

Both capacity and efficiency are independent qualities affecting production. Capacity may be associated with high or low efficiency. A harvesting machine of relatively low capacity, if operated at a high level of efficiency, may outproduce a machine of high capacity operated at a low level of efficiency.

In regard to forest operations it is necessary to recognize the difference between the two classical forms of forestry, namely exploitation and sustained yield forestry, with forests of all ages. The difference is that in sustained yield forestry the limiting factor is the forest resource, whereas in exploitation forestry, the forest resource, for all practical purposes, is unlimited and shortages exist in some other factor such as capital, labour, or markets.

Exploitation forestry has the objective to maximize the profit per unit of production. This objective is relevant when the forest resource is unlimited. When the forest resource is in short supply, sustained yield forestry with the objective of maximizing the profit per unit area of forest takes over.

In most textbooks on forest management little or no reference is made to harvesting technologies or the role of timber harvesting in forest management. On the other hand available textbooks on harvesting define this relationship clearly. For example in "Harvesting Timber Crops" (Wackerman et al, 1966) the authors state: "The harvesting of timber crops according to correct forestry methods may logically be considered the principal operation in forestry in that the culmination of all forestry effort is a merchantable tree for service to mankind. Harvesting therefore is one of the most important operations in forestry, entirely aside from the fact that it provides the income necessary to carry on the whole forestry business."

Tools used in forestry have changed with the times, but the end to which they are put has pretty much stayed the same. This is true in many areas, but is not universal.

Keays (1976) described this very well "In spite of the fact that methods and machines have changed beyond recognition within the past hundred years, the general principle of forest harvesting has remained the same. Forest harvesting in the past and in the present with few exceptions has involved the removal of the clear sound bole of the preferred wood species. Within this general principle there has been a steady general shift toward a more complete utilization of the forest. Stumps are cut lower and the top diameter is smaller. The concept of what constitutes a merchantable bole has changed and today it includes trees that would once have been culled because of size, defect or decay. Among the softwoods, trees are being used that were once considered to be weed species, and there has been a steady increase in the use of hardwoods. However, in spite of changes in utilization standards, the basic principle of harvesting the forest in terms of the sound bole of the preferred species has not changed."

Forests in the 17th and 18th centuries were used to supply building timbers and planks, wood for furniture and there was a considerable emphasis on supplying timbers of all kinds for the navies of Europe. Trees were felled and cut into logs and extracted from the forests. Although saws were known, particularly for sawing boards, either as pitsaws or water-powered sawmills, trees were felled by axe. Crosscut saws however were used to cut tree lengths into logs.

Until the beginning of the 20th century forests were exploited for saw timber and large wood products, poles, piling etc. Forest trees grew under natural conditions to large sizes and the use for such trees changed little with time. A certain stability of forest management and forest utilization developed. This situation changed as the uses to which wood was put changed from lumber and large wood products to pulps and papers, fibre and wood-particle-based products as well as energy.

2. OPERATIONAL EFFICIENCY

Operational efficiency is defined as the efficient utilization and economical management of the resource (forest). The aim of operational efficiency with which this book is concerned is to economize the human activities in forestry. Human inputs or activities are manipulated to meet certain goals or objectives.

The forest resource is a living biological system and it can be improved, retained, decreased or even destroyed by human activities. The forest is both a resource and a product and no clear-cut distinction can be made between the management of the resource and utilization of its yield.

Osmaston (1968) described the early human inputs as "Until he invented tools with which to cut and remove trees and until he discovered how to use fire to burn and destroy forests, man had no more impact upon forests than any other animal. Even then man's impact on forests was negligible until his command of his environment so developed that his population increased materially and at the same time his wants diversified."

In a rapidly changing and industrializing world the timber industry in a small country can only hope to maintain a relatively high rising standard of living by specializing in the production of goods it can produce more efficiently than its competitors in world markets.

It has been pointed out by Westoby (1966) that there are many reasons why investment in forestry and forest industries is attractive to lesser developed countries. Forest industries are diverse in range, economic size, capital requirements and sometimes in technological flexibility. Because most of the products are intermediate goods, this means that forest industry development has a multiplier contribution to make to overall economic growth.

The various stages of economic growth in labour-surplus countries and the structural changes required in the economy so that the re-allocation of the resources will lead to increased productivity have been described by Zivnuska (1966). Agricultural productivity must be increased so that the hidden reserves of under-employed farm labour can be released to the industrial sector. Simultaneously industrial employment opportunities might be developed so that this released labour force can be absorbed. In this context the fact that forest industries are typically rural industries becomes highly significant. Labour released from agriculture can be shifted more readily to timber harvesting than it could be to more highly urbanized and mechanized industries. The labour skills themselves are generally more readily adaptable to the requirements of logging. Operations such as logging can initially be conducted with high labour inputs and low investment in capital equipment until such time as labour is no longer in excess and higher degrees of mechanization can be introduced.

The expanded forest operations will convert the fixed capital in the form of standing trees to a more liquid form, cash, which can be directed to various sectors of the economy as needed. This capital can be invested in processing plants which can sustain a rising value productivity. It is apparent that the forest industries sector of the economy represents, to a large degree, an intermediate stage between the agriculture and general industrial sectors of the economy.

In considering new timber harvesting methods and machines it should be kept in mind that poverty is a direct cause of conservatism. If there is no margin to spare, there is no margin for risk. Since there is a measure of risk in anything that is untried, the tendency is to stick with the methods that have been proven and that have justified themselves to date. Usually any change is regarded with uneasiness. This explains to a large degree the conservatism that exists in subsistence agriculture the world over. It is also true for labour-intensive, timber-harvesting operations where there is not sufficient capital or margin with which to introduce new techniques and equipment. However to oppose technological change is to pass up the chance to eliminate heavy and distasteful toil. It means giving up any prospect of improving living standards (Silversides, 1968).

It is essential, however, that a certain level of technical activity must exist in a society for the successful application of technology to take place (Rybczynski, 1983). This fact was very evident in the forest operations in eastern Canada when tree harvesting and processing machines were first introduced. The machine operators and mechanics at that time were familiar with chainsaws, tractors, and automotive equipment. The new machines were, because of their very nature, more sophisticated with complex hydraulic and electronic circuitry. For example there was an almost quantum leap from low pressure tractor hydraulic systems to those of aircraft pressures (from 140 kgf/cm^2 to 350 kgf/cm^2) with the accompanying need for precision and absolute cleanliness that did not formerly exist. In this instance the cart arrived before the horse. The technology preceded the presence of the required technical competence. It is safe to say that most of the problems encountered in the introduction of new tree harvesting and processing machines have resulted from this situation. Current difficulties in the mechanization of harvesting operations in lesser developed countries are due in large part to this condition.

3. WORLD FOREST RESOURCES

It is necessary to define the concept "forest" in this text as the word means different things to different people. The Food and Agriculture Organization of the United Nations in its classification makes a distinction between a "closed forest" defined as having a crown cover of more than 19 percent and "open woodlands" as having a crown cover of 5-19 percent. Savannah forest types would be considered open woodlands while the closed forest types could be of possible interest for industrial forestry.

Knowledge as to the world's forest resources is really very limited and it is estimated that only about 40 percent of the forest area has been inventoried. The total gross volume in the world's closed forest is estimated to be approximately 300 billion m^3. Approximately 20 percent of the world's total land area is covered by closed forest and another 10 percent by open woodlands. Coniferous species make up one third of the total volume but only a small volume of conifers is to be found in developing countries. About one half of the total gross volume in the world is predominantly tropical hardwoods (Jullander and Stockman, 1978).

Table 1A. World Forest Area - 1980 (FAO, 1985)

Regions	Coniferous Total	Coniferous Closed	Non-coniferous Total	Non-coniferous Closed	Total forest area	Forest as % of land area (%)
		million	hectare			
Africa	13,9	7,8	730,0	228,1	743,7	24,5
N. and C. America	533,1	312,8	274,0	218,8	807,1	36,8
S. America	20,5	19,5	894,5	647,5	915,0	51,4
Asia	102,2	88,6	366,0	275,3	468,2	18,5
Europe	98,0	88,4	60,9	48,6	158,9	30,5
Oceania	16,0	12,1	283,0	211,3	298,9	27,9
USSR	699,9	593,7	248,7	197,9	928,6	41,5
World	1483,9	1120,8	2858,9	1827,5	4342,5	32,3

Forests grow throughout the world under a wide range of topographic and climatic conditions, from sea level swamps to high mountain slopes, and from the equator almost to the poles. Forests regulate climate and watersheds, control soil erosion, form centres of recreation and preserve wildlife. The rational utilization of forest products is essential for the preservation of forests. The major goal of forestry is timber crops in whatever form they may be used. The culmination of all forestry effort is the mature tree, whose value can only be realized when it is harvested. For this reason the techniques of timber harvesting and the economics of this operation are of great import in the development and maintenance of forest-based industries around the world.

Table 1A gives a breakdown of the world's forest resources, both "closed" and "open" and coniferous and non-coniferous and shows the percentage of the earth's land area that is forested.

Table 1B. Annual rates of deforestation and plantation in tropical regions, 1981-1985

Region	Annual area cutover Closed forest	Open woodland	Total	Annual planting	Ratio: planting/deforestation
	000 hectare				
Tropical America (23 countries)	4339	1272	5611	535	1:10,5
Tropical Africa (38 countries)	1331	2345	3676	126	1:29
Tropical Asia (15 countries)	1826	190	2016	438	1: 4,5
Total (76 countries)	7496	3807	11303	1099	1:10

Source: Lanly, J.P., 1982. Tropical Forest Resources, FAO Forestry Paper 30, Rome.

FAO estimates that $7292 \cdot 10^3$ ha of the forest resources are lost every year (Table 1B) of which $1480 \cdot 10^3$ ha are in Brazil, $800 \cdot 10^3$ ha in Columbia, $500 \cdot 10^3$ ha in Mexico, $333 \cdot 10^3$ ha in Thailand, $310 \cdot 10^3$ ha in the Ivory Coast and $285 \cdot 10^3$ ha in Nigeria.

An attempt is made in Table 1C to throw some light on the location and properties of the world's forests, based upon information from various sources. Where the forest resource is indicated as inoperable it means that it cannot be operated economically at present. The inoperability or inaccessibility of a forest will vary with the technology at the time of description. Detailed knowledge of marginal or submarginal resources is not very good and the statistics should be regarded as qualified estimates.

With regard to the inoperable resources in some of the regions:

1. The $190 \cdot 10^6$ ha in North America consist largely of the coniferous woodlands in Northern Canada and Alaska, close to the Arctic tree line, too remote from markets, with small-sized trees, low stocking, slow growth rates and poor

Table 1C. World Forest Resources - Location and Accessibility

Region	Total	Coniferous	Closed forest Area, million hectare Man-made forests	Unexploited	Inoperable	Inventoried %	Total volume billion m^3 with bark	Open woodland Area million ha	Total volume billion m^3 with bark(5)
Developed regions									
North America	630	(400)	(11)	(310)	190 (3)	78	54	- (4)	-
Western Europe	112	60	(26)	-	-	68	9,2	29	(0 7)
Oceania	49	11	1,2	20 (2)	20	57	2,0	102	(2[illegible]
Other market economies	26	11	10,2	1	1	100	2,2	3	(0 1)
Eastern Europe, USSR	793	568	(16+)	(540)	(440)(3)	42	77,2	115	(5 7)
Developing countries									
Africa	190	3	2	140	40	13	-	600	
Latin America	680	37	3,6	(540)(1)	250	15	-	280	
Near East	10	4	0,2	-	-	43	-	13	
Far East	270	11	3,2	125 (2)	85	30	-	(50)	
Other market economies	39	0	0	37 (2)	30	17	-	4	
Centrally planned	130	60	30-60	50 (2)	23	-	-	-	
Total	2929 (6)	1165	-	-		(42)	300	(1500)	40

Sources: In the main Persson 1974 and 1977, FAO 1976, FAO/RLAT 1976 FAO/ECE 1976.
(1) Shows mainly inaccessible forests. (2) Shows forest not in use. (3) Not suitable for regular harvest.
(4) Probably included in closed forest area. (5) Bark may sometimes be excluded. (6) Total land area is 13.4 billion hectare.

soils for road construction and operating logging machinery.
2. The same applies to the $440 \cdot 10^6$ ha in the USSR.
3. The inoperable resource in developing countries, estimated to be $428 \cdot 10^6$ ha, 60% of it in Latin America (e.g. the Amazon basin) consists of tropical forests with low stocking of presently merchantable species, remote from markets, with poor soils and difficult climate. The large size of the trees requires very large, heavy equipment and their exploitation results in environmental problems.
4. Some of the inoperable forest resources are located on extremely steep terrain.

The open woodlands, $1500 \cdot 10^6$ ha are mostly arid savannah forest, often very important for the local people for their supply of firewood and also fodder for their cattle. These areas will probably never contribute to industrial use.

The total production of industrial wood in 1984 was approximately $1,5 \cdot 10^9 m^3$. About the same quantity was used for fuelwood, mainly in developing countries. This makes wood the single largest commodity for man's need.

The removals from the forests are often computed in unit volume per unit forest area and related to the actual or potential growth. It is evident that such assessments of a global, regional and even national supply/demand balance can be very misleading. There are many constraints in allowing a surplus in one area to compensate for a deficit in another. Most wood assortments cannot be moved long distances and species and quality as well as inaccessibility and infrastructure may severely restrict the use of a forest resource. Great caution must be observed in studies of wood supply.

Note: In accordance with international convention, the comma is used as the decimal sign throughout the book. The decimal comma is accepted as correct even in countries where the decimal point is common practice.

4. FOREST MECHANIZATION

In agriculture, harvesting was the crucial operation in the production of grain. The mechanical reaper was introduced in the United States between 1830 and 1860. Harvesting is the crucial operation in the production of sawlogs and pulpwood and the mechanical tree harvester was introduced in the 1960's. It was the American Civil War which ultimately provided the impetus for change from manual labour to mechanical power in agriculture in the United States and it was World War II which provided the impetus in the mechanization of logging in North America and northern Europe.

It has been thought that machines developed for the logging industry could be considered as simple variations of agricultural equipment. In fact the logging industry is quite distinct from any other and has many unique problems. The only similarity between agriculture and forestry is that they are both concerned with the planting, growing and harvesting of plants (Janosi et al, 1970).

However it would appear that harvesting machines and systems, regardless of the crop, pass through analogous stages in their development. The developments in the mechanization of agriculture preceded those in forestry. The forest environment, on the whole was too hostile. Forest land in most countries is non-agricultural land, for various reasons. The terrain may be too rocky, too wet, too steep or otherwise unsuitable for agricultural purposes.

There were three major developments in agriculture in Canada in the 1930's and 1940's. The first was putting rubber tires on farm machines. Rubber tires paved the way for lifting farm operations above the walking gait of man and horse. The second innovation was the fitting of implements to power plants. Early tractors pulled implements, modern tractors carry them. The third innovation was the mounting of power units on the machine, i.e. self-propelled combines and harvesters of various types.

A similar progression of development took place in timber harvesting. Internal combustion engines brought light into the woods. Crawler tractor bulldozers permitted economic road construction. Initially tractors and trucks were used to drag loads. The articulated frame-steered wheeled vehicle, whether skidder or forwarder, changed timber harvesting from a seasonal operation, the tempo of which was controlled by that of animal power, to a year-round operation paced by tireless machines. The third innovation was the development of tree harvesters which felled and processed trees entirely by machine. As in agriculture many variations of such harvesters were developed.

4.1 HARVESTING SYSTEMS

Logging methods are established methods or ways of harvesting trees. The term "method" applies to the manner in which the operation is conducted as well as the form of the product. It could be the shortwood method in which trees are felled, limbed and cut into short lengths or bolts which are usually handled manually and carried and placed in small piles along prepared roads or trails in the forest; or the full-tree method where trees are felled and transported in their entirety to be limbed and further processed at roadside.

A logging system varies from a logging method in that the various operations in reducing a standing tree into logs or bolts are so related that each is sequentially dependent upon the other in time and space so that any interference at any stage of production destroys the sequence and production falters or comes to a halt. In any system each step affects the subsequent step and in turn is affected by the preceding step.

Many analyses have been made of timber harvesting systems in an effort to establish which is the most economic for a given situation. Mellgren (1978) stated that there was no significant difference in logging costs or in investment per unit of production between various machine systems as long as proper training and servicing were provided to all. His one stated exception was the multi-tree system.

Boyd and Novak (1977) found that when analyzing tree-length, full-tree and shortwood systems that there was no apparent conceptual advantage of importance for any one of these major system types. They also noted that the system with the weakest concept, if operated at maximum performance achievement (efficiency) and if it were well designed and engineered it could produce wood at lower cost than the conceptually best system operated with its performance factors at the minimum achievement level.

It is argued by some that it is best to make up a uniform package of wood in the forest as close to the stump as possible, with the elimination of all unusable material from the tree, and to handle this optimum-size package as such from there to the final destination. It is argued that it is uneconomic to transport any unusable material as it adds to cost not only in transport but in disposing of it later. Others claim that only the minimum amount of work should be done on a tree at the stump and that all processing should be carried out at a central point with highly productive machines. These are the two extremes with intermediate procedures in between, such as the tree-length system.

There does not appear to be any one system of logging that will have universal acceptance. The physical and economic variables existant are such that no one system can operate under all circumstances at optimum capacity and cost. Only as each logging situation is analyzed in some detail is it possible to determine which logging system should be used. Normally the forest stand and terrain characteristics, which will be discussed later, are governing.

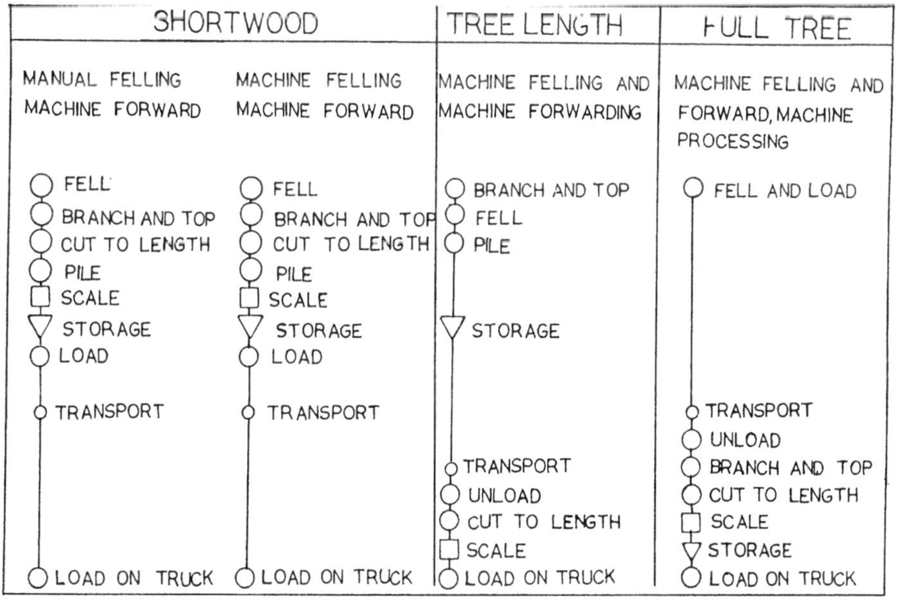

Figure 1. Flow charts of various harvesting systems.

A number of basic attitudes have been expressed but it is difficult if not impossible to quantify their influence on productivity and cost. Some of these attitudes are:

- it is basic to the principle of mass production that raw material be transported to the work place. The worker on the production line does not have to seek out items with which he has to work, but they are brought to him, to maximize his output. In logging, this could mean working out of stockpiles with a highly productive machine such as a full-tree processor or roadside chipper.

- it is considered poor economics to transport unusable material if it can possibly be avoided. Therefore the full-tree system which involves the transport of unusable limbs and tree tops from stump to roadside is suspect physically if not economically.

- is it right to have a very costly harvester-processor travelling through a forest with stands of variable density with a varying content of merchantable trees and over variable terrain, seeking out and harvesting and processing trees of small average volume?

- it would appear logically correct to produce a product of final specifications as close to the stump as possible and to only transport material that is usable and of value.

If trees or logs must be subjected to examination and/or measurement then the harvesting system has a constraint on its output, imposed by such examination or measurement. If the product, for example, is hardwood sawlogs the operation is normally carried out on the basis of quality product rather than quantity of production and production is usually below that for softwood sawtimber.

When logging methods were labour intensive, it was possible for the forest worker to cull defective timber and to grade the material he put into skidways or piles early in the harvesting process. The quality standards set by the purchaser acted as a silent foreman and prohibited the inclusion of defective wood to the degree that the workers would not be paid for such wood if included in their piles.

Mellgren (1978) has outlined some of the various arguments for and against the shortwood, tree-length and full-tree systems.

Shortwood
- minimum rehandling, sawlogs directly to sawmill, pulpwood directly to pulpmill
- compact loads
- clean wood
- better sorting
- smaller woodyards
- better recovery
- requires complex multi-function machines to operate in forest stand.

Tree Length
- few pieces to handle
- requires costly sorting and slashing
- bulky truck loads
- high breakage losses

Full Tree
- few and simple machines at stump area
- higher forwarding costs
- slash removal problems at landing.

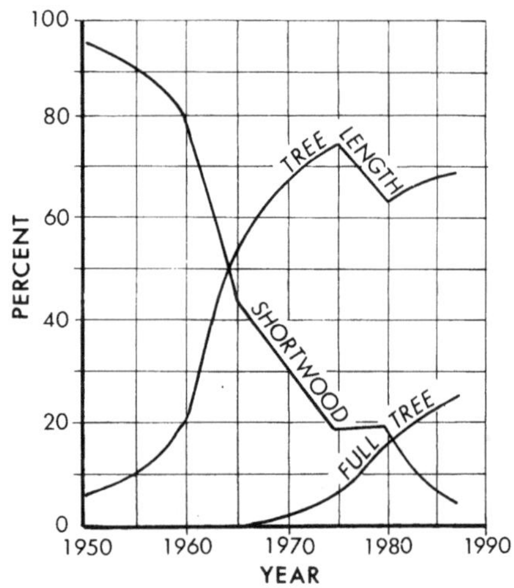

Figure 2. Logging methods of limit wood in eastern Canada, 1950-1987, in percent (%) of total harvested volume.

Some of the above factors truly impact upon the productivity and cost of a system, others have really little effect.

With the advent of mechanization of logging practices it has been found impractical to attempt to cull out defective wood during production operations. If trees are felled and collected by machine, are handled and cut to length in multiples there is little or no opportunity to be selective with regard to defective trees. As a result of mechanization of harvesting a good deal more defect or cull is being found, and in many instances, accepted, than occurred when logging was labour intensive.

The final product will dictate to a large degree the harvesting system used in any particular instance. It would not be practical to use some forms of the so-called shortwood system producing two metre length wood for the production of sawlogs for random length lumber. On the other hand the production of tree length or full trees to landing or roadside where sawtimber, pulpwood and other products could be sorted out lends itself to the selection of final product.

4.1.1 SHORTWOOD. When harvesting pulpwood was labour intensive shortwood was cut and piled manually. A common sight on such operations was a straight cleared strip road bordered on each side with neat piles of pulpwood. The piles, after measurement for payment to the cutter, the contractor and the forest owner, were broken down and loaded manually onto either wagon or sleigh. An early development in mechanization was of ways and means to move the piles as a whole. It wasn't until the hydraulically operated knuckle boom loader with grapple was introduced that the system changed. It was accepted that the only reason wood was put up in neat piles was to measure it and to move it as a package. By simply bunching wood in small piles to be handled by grapple rather than piled in neat piles it was found that the output per forest worker was increased about 30 percent. Counting the pieces of wood and applying conversion tables took the place of having to measure the neat piles. The piling operation itself added nothing to the manufacture of the shortwood.

The economic implementation of the shortwood system in the boreal forest has depended upon the development of suitable forwarders equipped with grapple loaders. Labour-intensive manual felling and crosscutting of trees have continued in some regions but are being replaced by a variety of mechanical tree harvesters capable of felling, limbing, crosscutting and piling bolts and logs into bunches ready for forwarding.

The wood thus produced is transported out of the forest by self-loading forwarders. Approximately 90 percent of the timber harvest in Sweden is transported by such forwarders. Only 5 percent of the pulpwood harvest in eastern Canada is produced by the shortwood system and forwarded from the forest in this form.

Current forwarders have articulated-frame steering with oscillating frames or axle. A few are equipped with four large wheels but six- and eight-wheeled units, the latter equipped with double bogies, are more common. The machines vary in power from 35-145 kW and in load capacity from 6-15 t.

Forwarders are used almost exclusively to transport shortwood although some tree-length forwarders have been introduced. Skidders are used primarily to transport tree lengths and full trees but may also be used to carry bundles of shortwood. Forwarders and skidders are compared in the following list.

Table 2. Relative charcteristics of rubber-tired forwarders and skidders

Forwarder	Skidder
1. Articulated frame-steered rubber-tired chassis.	- Articulated frame-steered rubber-tired chassis.
2. Operated by one man.	- Operated by one man.
3. Loads and unloads self with a hydraulically operated grapple loader. Can pile logs in high stacks.	- With cable-equipped skidder the operator must dismount to attach and detach loads. - With grapple-equipped skidder the operator picks up and deposits load from seat.
4. Operation of unit is independent of the preceding and following operations.	- Operation of unit is independent of the preceding and following operations.
5. Load is wholly carried on machine.	- Load is dragged behind machine.
6. Machine is highly mobile but when loaded has a high centre of gravity.	- Machine is highly mobile, centre of gravity remains low.
7. If immobilized, machine can unload itself.	- Load can be readily dropped and later winched in, if machine must extricate itself.
8. Forwarder normally transports larger loads than skidders but at slower speeds.	- Skidder transports smaller load than forwarder, but at higher speeds.
9. Machine and load has one point of articulation, usually at or near centre of machine.	- Machine and load have two points of articulation, at centre of machine and at point of load attachment under fairlead at rear of unit.
10. Length of machine and load is fixed.	- Length of machine and load varies with length of trees in load.

4.1.2 TREE LENGTH. Tree-length skidding became a major harvesting method in North America with the introduction of the articulated frame-steered skidder. Tree lengths had long been skidded with horses. However animal power became a limiting factor on production and the logging industry turned to tireless machines with an almost infinite variety in power and size to replace the horse. Although the machines cost up to 20 times that of a team of horses, they resulted in increases in productivity of up to 1500 percent. Many variations were introduced to increase output of this method, the limitation being the number of chokers possible to use when harvesting trees of small dimension. Different felling patterns were developed to increase load size, i.e. to permit two or more tree lengths to be attached by one choker; choking butt end and top end were tried to establish the effect upon skidder load and speed of travel etc.

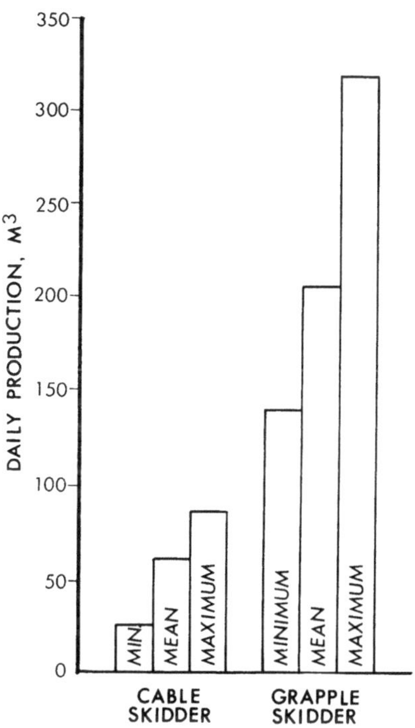

Figure 3. Comparison of productivity of cable and grapple skidders based upon 8-h day (Sarles and Luppold, 1986).

Tree-length logging started to come into its own in eastern Canada in the early 1950's. It increased rapidly over the succeeding twenty years at the expense of the shortwood system which has declined almost steadily since 1950. There was a short period 1975-1980 when there was a sharp decline in the tree-length method and a slight increase in the shortwood method. This was the result of the successful introduction of the Koehring Shortwood Harvester between the years 1970 and 1977. It has since passed from the scene and shortwood production has been surpassed by that of full trees (Figure 2). The size and power of skidders have tended to increase with experience as larger loads and higher speeds produce lower cost wood. The greatest increase in productivity developed with the introduction of the feller buncher machines and the skidders equipped with grapples. This development reduced the impact of tree size which was critical when operating a cable-equipped skidder with chokers. Because of the relatively fixed number of chokers it is possible to use, the maximum load depends upon tree size. The skidders were often moving loads far below this capacity when operating in small-size timber. With a grapple skidder, the grapple capacity controlled the size of load and this in turn could be produced by the feller buncher. In theory at least the grapple skidder could operate with an optimum load each trip (Figure 3).

A recent development in tree-length skidding is the so-called clambunk skidder. In the early development of tree-length skidding a number of attempts were made to mount a knuckle boom loader behind the cab of a standard short wheelbase skidder and to have an upside down grapple fitted on the rear of the skidder, in place of an arch and winch. The upside down grapple or clambunk was to be loaded to capacity with tree lengths, butts foremost, for skidding. Most trials were unsuccessful due to poor weight distribution and manoeuverability until a chassis was designed specifically for this application. The wheel base was lengthened and the rear large single wheels were replaced with tandem bogie wheels which permitted a lower centre of gravity, and the clambunk was redesigned. The first clambunk skidders appeared on operations in Scandinavia and spread from there to North America. The clambunk is slower than a standard wheeled skidder but is capable of moving large loads (13 m³) compared to perhaps 4-6 m³ with a wheeled grapple skidder. Because making up the load forms the largest part of the cycle time this unit has proven most applicable on long skids (+ 300 m) and where road construction costs are high (Figure 4).

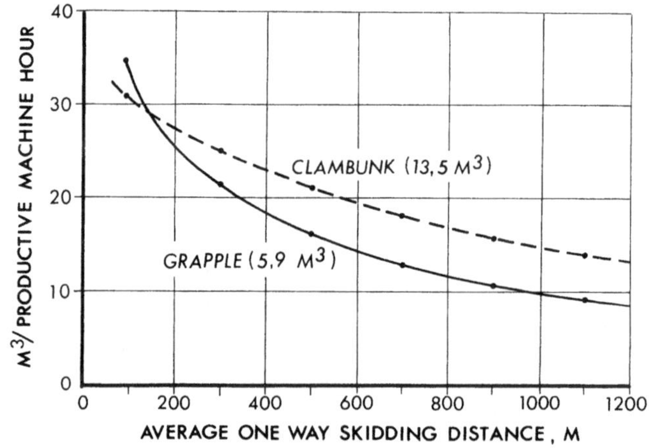

Figure 4. Production in off-road transport, grapple versus clambunk skidder. Average load: grapple skidder 5,9 m³, clambunk skidder 13,5 m³ (Mellgren 1987).

An apparent advantage of self-loading clambunk skidders is their capability to optimize and control machine capacity during each operating cycle similar to self-loading forwarders. The optimization of the payload size with grapple skidders is not as readily possible as it is dependent upon the average bunch size and bunch alignment as they occur in the forest.

A number of studies have established that the combined effect of skidding distance and payload volume is the principle factor determining the transport system productivity (Hallett and McCraw, 1970) (Bennett, Winer and Bartholomew, 1965) (Jiles and Lehman, 1960). Payload optimization on each trip is critical with high-capacity transporters over long skidding distances.

The rapid acceptance of the tree-length system resulted in large degree as a result of the development of efficient multi-stem high-capacity roadside slashers to reduce the tree lengths to prescribed lengths, as well as the development of efficient transporters.

4.1.3 FULL TREE. By full-tree harvesting is meant the severance of a tree from its stump and its transport to roadside or landing with no further work done on it in the stump area. Figure 1 clearly shows the concentration of the various activities both inside and outside the stand. It is argued that production would be higher and costs lower if trees were processed, that is, limbed and topped, perhaps cut to length, in concentrations at roadside or landing rather than individually in the stump area. The principle is that an absolute minimum of effort be expended in the forest and a maximum where trees are concentrated and where conditions of work can be better controlled.

Full-tree harvesting, in eastern Canada primarily, has grown steadily since the late 1960's until in 1987 it is used to produce 25 percent of the pulpwood in this region. As early as 1950 the U.S.S.R. undertook to establish full tree as the standard harvesting method. In 1953 under the auspices of the Canadian Pulp and Paper Association a study was undertaken by B.J. McColl and W.A.E. Pepler and published as "The Development of Mechanical Pulpwood Logging Methods for Eastern Canada." A major conclusion in this report was "the full unbranched tree method provides the greatest possibility of successful mechanization" (McColl and Pepler, 1953). In 1954 Alexander Koroleff after studying developments in the U.S.S.R. generated further interest in this logging method when his book "Full-tree Logging - A Challenge to Research" was published by the Pulp and Paper Research Institute of Canada (Koroleff, 1954).

Sporadic trials were undertaken since the mid 1950's but the method did not really spread until the late 1970's. The success of the full-tree method was dependent upon the development of machines to remove limbs and to cut the tree boles into prescribed lengths. This equipment appeared in the late 1970's and early 1980's.

Full trees can be produced with feller-buncher machines. They can be transported to roadside by choker, grapple or clambunk skidders. Because of the additional weight of limbs and tops, and the nature of the branches it has been found that it requires twice as much power to skid full trees as it does the same trees as tree lengths (Garlicki and Calvert, 1969). A new development is that of the feller forwarder, a machine which fells trees, loads them onto itself and carries its load of trees (forwards) to the roadside.

Full-tree harvesting has raised as many questions at it has solved. It does appear under many conditions to be the most economic system of harvesting coniferous stands. However there is concern that the removal of all the above-ground portions of trees may result in the depletion of nutrients from the site. Studies have shown that all species have a similar pattern of nutrient accumulation in the various biomass components. All species show higher concentrations of nutrients in the small biomass components. This appears due in part to higher concentrations of bark in these components. This is illustrated in Table 3 (Chatarpaul et al, 1984).

Full-tree harvesting removes all limbs and tops from the cutover area except those broken off during felling and skidding operations; up to 80 percent of the area may be left with no cover of limbs and branches to present a fire hazard. All that remain standing are unmerchantable and dead trees. This mechanized logging system should result in a minimum fire hazard within the cutover area. Instead of being scattered throughout the cutover area the limbs and tops are concentrated on landings at the roadside or are hauled away to the mill along with the bole wood.

Table 3. Nutrients (%) and bark/wood ratios, averaged for all species (1)

	- Ca -		- K -		- Mg -		- N -		Ratio
Component	Wood	Bark	Wood	Bark	Wood	Bark	Wood	Bark	Bark/Wood
Branches 0,5-2,5 cm	0,215	1,127	0,095	0,272	0,038	0,087	0,153	0,538	0,329
Branches >2,5 cm	0,212	1,479	0,106	0,219	0,031	0,078	0,135	0,438	0,220
Main Stem <9,2 cm	0,158	0,154	0,101	0,285	0,034	0,080	0,115	0,499	0,198
Main Stem >9,2 cm	0,172	0,894	0,100	0,217	0,028	0,063	0,074	0,357	0,130

(1) Boreal mixedwood stand containing coniferous and deciduous species.

Much of the nitrogen on site in the boreal forests is contained in the foliage and fine branches and twigs and these yield little additional biomass in the whole-tree harvest.

4.1.4 UNLIMBED TREE SECTIONS. This is a variation of the full-tree system in that the full trees are cut to convenient lengths for handling and transport.

After the declining use of wood as fuel, due to displacement by cheap fossil fuels, biomass is again attracting attention and interest in many countries. In Scandinavian countries a hybrid method between shortwood and full-tree logging has developed for thinning operations in young stands. The trees are felled with a feller forwarder equipped with a knuckle boom with a long reach. The trees, after felling, are cut to a convenient length, about 4-5 m, by the feller buncher and transported to roadside or landing by the feller forwarder or a forwader. The unlimbed tree sections are then hauled by truck to a pulpmill and put through a drum delimber debarker to produce two products: debarked pulpwood and fuel (bark, needles, twigs, tops). The merit of this system is that it reduces the cost of thinning, as delimbing within the forest stand is eliminated and it results in a higher yield of biomass per unit area, both as fibre for pulping and as fuel.

Transport of unlimbed tree sections is more costly than that of bolts and logs due to low bulk density of the load but this is overcome in part by load compression devices. This system is described in more detail in Chapter 11.

4.1.5 UTILIZATION OF LOGGING RESIDUES. Logging residues, tops, branches, small trees below current utilization standards are being used in some countries for energy. On conventional logging operations the material is collected by a second pass over the logged area with a forwarder equipped with a suitable grapple. The material is moved to roadside and chipped or crushed and transported by truck to a woodburning installation and is usually used for heating schools, hospitals, or groups of houses. With full-tree logging a second pass is not required as the material to be chipped may already be accumulated at roadside. See Chapter 11.

4.1.6 IMPACT OF MARKETING ON LOGGING SYSTEMS. Roundwood is used for a large variety of end products and logs of special qualities can obtain very high prices in comparison with standard assortments of sawlogs and pulpwood. A private forest owner may be able to take advantage of this market to increase his revenues and may give preference to logging systems that give the best opportunities to do so. As there is a disadvantage to have too many systems in use at the same time, marketing situations may have a considerable impact on system dominance, especially in regions with a large share of farm forestry.

4.2 CUTTING

The tools used in timber harvesting have changed dramatically in recent years. Up until World War II hand tools dominated and it was not until the post-war period that tools became powered and machines capable of performing the various tasks in falling, crosscutting, delimbing and loading were introduced into the woods.

The changes in cutting technology are of particular interest. Felling and crosscutting are considered to be the two most arduous tasks in manually harvesting trees and much effort has been expended to lighten this burden on forest workers.

4.2.1 AXES. Initially the axe was the universal tool. Its design and use go back before recorded history and the use of metals. Excavation of Mesolithic sites (5000 B.C.) in Denmark has produced axes weighing a kilogram or more, made by flaking a heavy nodule of flint into an elongated implement with a sharp edge. These implements were fastened to handles made from elm or ash. Such axes could be resharpened by chipping off a large flake or flakes across the leading edge.

The axe in America was reshaped from its European ancestor which had changed little since late Gothic times (Giedion, 1948). Originally the axe heads were unbalanced and the handle was straight and round in shape. In America the axe was greatly modified with its curved cutting edge, heavier head counter-balanced by its poll which gave it better handling qualities, a curved handle which permitted it to be more easily guided and more forcibly swung, and oval in shape to prevent it twisting in the user's hands. The new designs rapidly replaced the older models brought from Europe. The poll added weight to the axe head and must have affected the dynamics of the axe swing and the energy it could transmit to the wood on impact (Widule et al, 1978) (Figure 5).

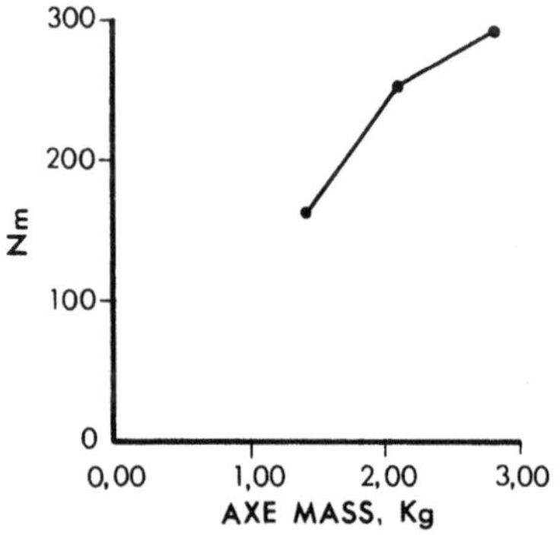

Figure 5. Kinetic energy in Nm generated in axes of different mass.

The use of the axe in chopping, whether felling trees or in crosscutting tree boles into logs is considered to be heavy work. The metabolic energy expended in chopping with an axe has been found to vary from 400-1400 kcal/h. Typically the heaviest work a young fit male can sustain for a prolonged period is 500 kcal/h and for the general population 400-425 kcal/h (Hertig, 1979).

Table 4. Energy comsumption in chopping (kcal/s)

		Moderate	Heavy	Very heavy
1. Normal rating of work level		0,08	0,13	0,18
2. Chopping with axe (Maleta and Sood, 1984)		-	0,13	-
3. Chopping with axe (Hertig, 1979)		-	0,11	0,39
4. Chopping with axe (Smith, 1980)				
Axe weight/kg	Blows/min			
1,25	20	0,09	-	-
1,25	35	-	0,16	
2,0	35	-	0,17	
1,25	50	-	-	0,32

4.2.2 SAWS. In eastern Canada and elsewhere the axe was the tool taken to the woods to fell and crosscut trees until approximately 1850 when the crosscut saw was introduced. At that time the steel industry had developed the technology to be able to roll thin steel sheets of the required strength to produce saws. Since then there has been a progression of

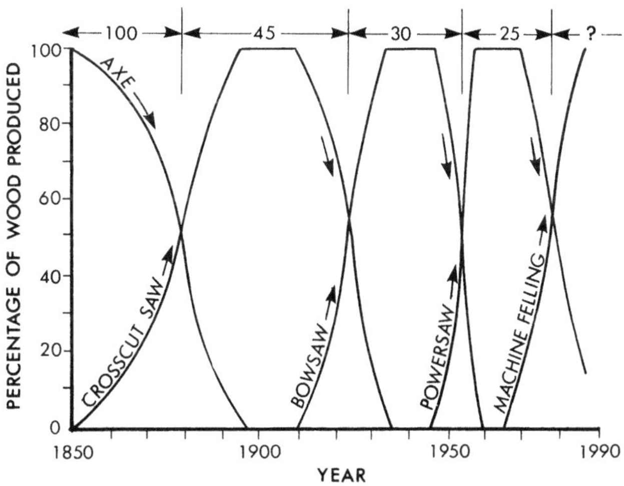

Figure 6. Pattern of substitution of cutting tools in eastern Canada.

tools as the crosscut saw was displaced by the bowsaw which was in turn displaced by the powersaw. The axe, crosscut saw and bowsaw are now almost museum exhibits as they have disappeared completely from the woods. The most recent development is the partial displacement of the powersaw by felling machines.

It is not anticipated that felling machines will totally displace powersaws as there are extensive areas where terrain conditions and tree size prohibit the use of felling machines, at least in their present configurations and in developing countries manual tools may still be the most economic choice, especially in the cutting of smaller trees.

The two-man crosscut saw was known to the Romans but didn't come into common usage in Europe until the 15th century. Records indicate that the crosscut saw was used in the United States for producing logs as early as 1635. At this time trees were axe felled and then sawn into desired log lengths. It wasn't until about 1880 that loggers began felling trees with the crosscut saw (Miller, 1978).

Miller (1978) describes the mechanism of cutting with a crosscut saw. The cutting, or pointed, teeth of the saw sever the fibres on each side of the kerf, or slot, the saw makes when cutting. The raker teeth, acting like a plane, peel off the cut fibres and collect them in the sawdust gullets, the spaces between the cutting teeth and the raker teeth, and carry them out of the cut. For large diameter trees, where considerable shavings accumulate with each stroke, a large gullet is necessary to remove the shavings to prevent the saw from binding (Figure 7).

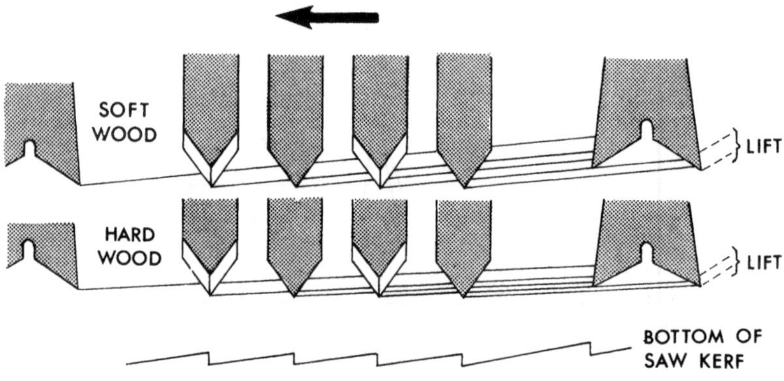

Figure 7. The function of a raker tooth saw in soft wood with large feed and in hard wood with lower feed, showing the adaption of the depth gauge setting to wood properties. Note that the rear point of the raker rides on uncut wood, causing a lift or compression on the wood.

The one-man bowsaw has been studied at some length, in relation to its efficiency in cutting. It was found that there was little difference in the efficiency of saws of different tooth design if energy consumption is used as the basis of measurement (Wakefield, 1932). It was found that the raker teeth were of advantage in that by the expenditure of additional energy, an increase in speed of cutting could be obtained with lower saw pressure. The most efficient pressure was found to be about 5 kg for V-tooth saws. Increases in pressure are accompanied by increased cutting speed though not in direct proportion.

The bowsaw was rapidly displaced by the powersaw in the post World War II period. The first patents granted for a motor chainsaw or powersaw were in 1850 in Germany and in 1858 in the United States. There were many attempts to produce saws light enough to be of practical use in the forest but the break through did not come until after World War II.

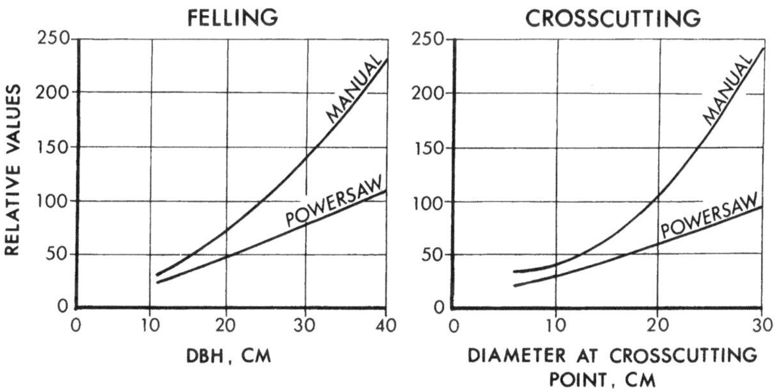

Figure 8. Relative values sawing manually with bowsaw and with powersaw.

The time lag between the conception of the idea of a powersaw and its successful innovation was almost a century. The technology of the times did not provide the essential conditions for its development. It was not until two developments which arose out of wartime activities, neither of which had any bearing on the saw design, made it all possible in a relatively short period of time. One development was in light metal technology (aluminum and magnesium) and the other was the development of light gasoline-powered air-cooled engine technology.

Physiological studies had shown that tree felling and cross-cutting are the two heaviest forms of work in the forest, requiring energy close to the upper limit of human physiological efficiency. Felling by handsaw was found to be the most energy-consuming operation (Sundberg, 1956). As soon as a powersaw of acceptable weight and reliability appeared it was rapidly adopted in the forest in Europe and North America. Figure 8 illustrates the advantage of the powersaw over manual felling and crosscutting.

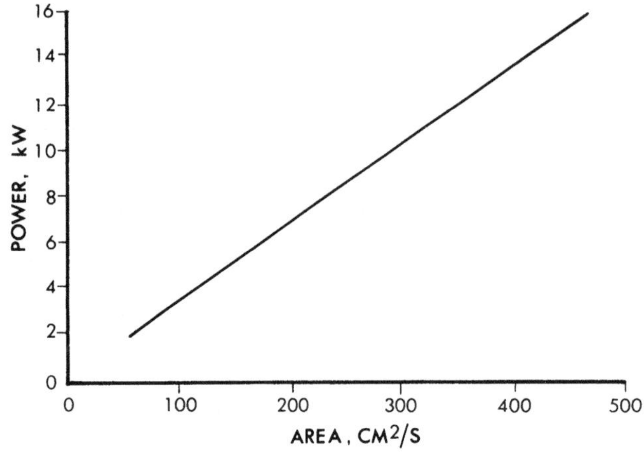

Figure 9. Sawn area of wood in cm^2/s as function of powersaw power in kW.

Figure 10 shows the time required to cut trees of varying diameters for different sawchain speeds (Palm et al, 1968).

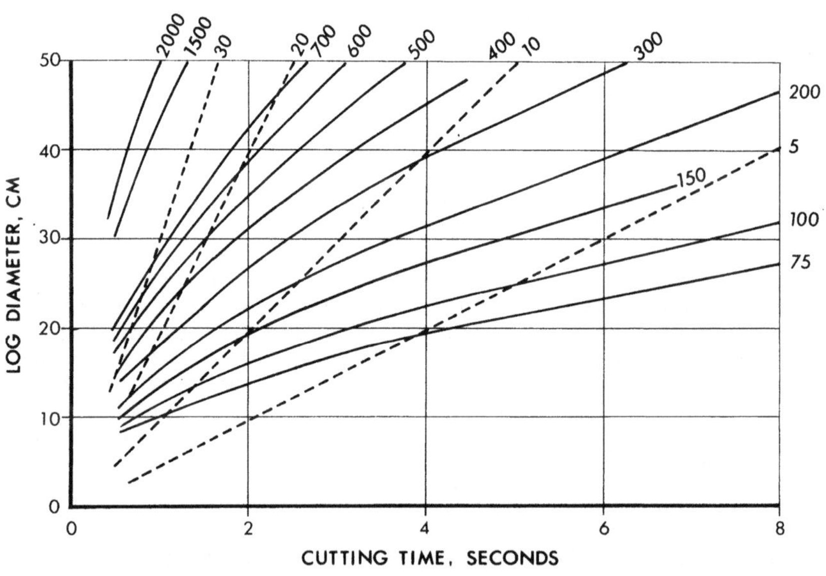

Figure 10. Cutting times for different log diameters:
full lines - for sawchains of various capacity, in cm^2/s, time proportional to d^2.
dotted lines - for various sawchain speeds, in m/s, time proportional to d.

Hazzard (1969) states that chainsaws, well maintained should average 1,7 m^2/min/kW and this is illustrated in Figure 9, expressed as cm^2/s.

4.2.3 SHEARS. In spite of the wide acceptance of the power-saw, its rapid development of a high state of reliability and its very major contribution to timber harvesting, the complexity of the chainsaw led to a search for simpler mechanisms and the concept of shearing wood developed. The cutting of wood with knives up until the late 1950's had been in the production of veneers and the wood was cut parallel to the grain into very thin sheets. The cutting of trees or logs with knives at right angles to their axis was a new departure. Experiments in felling trees with shears in North America goes back to 1945, used specifically for land clearing. The first reference found in the literature of the use of shears related to timber harvesting in North America was in 1954 in Louisiana, U.S.A. on the operations of the International Paper Company. The shear was mounted on a farm-type wheel tractor which was part of an experimental project to develop a mechanical tree-harvesting machine (Cline, 1967). In 1964 in the southeastern United States a sudden shortage of pulpwood developed and to increase the productivity of the existing labour force the felling of trees was mechanized. This was accomplished by mounting a hydraulically operated tree shear on the front end of a crawler tractor. The shear consisted of a single blade acting against an anvil. This particular configuration was not successful in stands of low density and small tree size but was widely accepted in regions which supported dense stands of large average tree size.

A major problem encountered in felling trees by shearing them from their stump is butt shatter, the splitting of the bole adjacent to the cut. This is of particular concern where the trees are utilized for the production of lumber.

In the action of shearing wood at right angles to the axis of the tree or log the fibres ahead of the edge of the shear blade fail, mostly by tensile forces with the blade edge acting to concentrate the stress. The force opposing the shearing action is made up of three components, the edge effect, the wedge effect and friction. The edge effect is that required to compress fibres ahead of the edge of the shear blade, the wedge effect is the force required to deflect these fibres after cutting, to a distance of half the blade thickness, and the force required to overcome friction on the sides of the blade (Johnston, 1967).

In studies carried out by Kempe (1964) experimental values for the force opposing shearing action were established and are of value as guides and can be considered as the orders of magnitude in which such forces lie. The edge force is considered to be approximately 50 kp/cm of the length of edge. The wedge force is of the order of magnitude of

100-160 kp/cm^2 of the projected surface of the shear blade (the surface of the blade x cosα, where α is one half the angle of the blade). The range is due to the edge angle being 100 kp/cm^2 for 30^0, 125 kp/cm^2 for 45^0, and 160 kp/cm^2 for 60^0. The specific friction force acting upon both sides of the shear blade is approximately 50 kp/cm^2.

When a blade or a wedge that is 3-5 cm thick at the back is forced into a tree at right angles to its axis it must displace a volume of wood equal to the volume of the blade. The stresses imposed upon the wood normally take two forms, intense and localized crushing next to the blade and longitudinal splitting to relieve the stresses set up in the tree. This situation appears to be inherent in shearing wood perpendicular to the grain. It is normally not of great concern if the tree is going to be used for pulpwood but can cause serious losses if it is to be used for lumber, veneer, or other products such as poles, piling, etc. Different attempts have been made to reduce the damage but none has proven altogether satisfactory.

Shears may have one blade acting against an anvil or two blades pivotting against each other with the blade edges acting as scissors, or two blades whose edges remain parallel to each other as they open and close. Blades to be used for felling trees have been dished or designed in a V so that through this blade geometry the stresses, set up while shearing is taking place, are directed into the stump to shatter it with minimum damage to the butt of the tree.

Single-blade shears result in directional felling of trees, as they tend to fall in line with the closing of the shear blade against its anvil. The wedging action of the blade forces the tree to fall in the direction of blade movement. Sometimes the shears are equipped with directional felling devices to ensure the direction of tree fall and to counteract the effects of lean or curvature of the tree. The tree is not supported above the shear blade so that as the blade cuts into the tree, the tree is free to bend at its hinge, thus tending to relieve some of the stresses involved.

It has been found that an increase in anvil width causes an increase in cutting force and specific energy required. This is because the increased width of anvil reduces the amount of bending of the tree or opening up as the blade advances. Opening up of the cut reduces the friction forces acting on the surfaces of the shear blade and it also causes tensile stress in the fibres immediately ahead of the cutting edge of the shear. With a single shear blade, cutting force increases to a maximum at a penetration of about two thirds the tree or log diameter after which it decreases at about the same rate until completion of the cut. With two shearing blades the cutting force increases steadily up to the completion of the cut, a little before either knife penetrates to half the diameter (Johnston, 1967).

With double-bladed shears, mounted on a grapple to hold the tree while shearing, bending stresses may be set up as the tree is not free to fall and, in operation, damage from this is common. Sometimes machine operators using the grapple will bend the tree during the shearing operation in an attempt to ease or speed removal from the stump. It has been found that wood damage caused by the blades alone has an average length of about twice the blade thickness but when bending forces are imposed on the tree, splitting damage can be up to a metre in length (Johnston and St. Laurent, 1974). The grapple holding the tree and the shear blades should be designed to permit some freedom of movement during the shearing operation.

With single-bladed shears, one attempt to reduce damage to the wood has been to groove the upper surface of the shear blade to not only reduce the friction between wood and its surface but also to generate severe local crushing of the butt surface of the tree to absorb all the stresses there, and thus to reduce or eliminate long splits.

McLauchlan and Kusec (1975) in studying the shearing of wood found that high stresses developed in the wood adjacent to the surface of the shear blade and as the blade advanced, release of these stresses was accompanied by explosive sounds and attendant fracturing of wood. A shear blade with triangular ribs on its upper surface (Figure 11) was tested. The purpose was to interrupt the flat surface of the blade and to limit the size of the stressed areas. The sharpened leading edges of the ribs separate and compress the wood fibres perpendicular to the grain. Wood is comparatively weak in compression parallel to the grain and it proved possible to create spaces and permit entry of the ribs without creating excessive stresses. The ribs reinforce the blade and permit a thinner blade to be used safely, and this in turn tended to cause less damage to the wood. Tests carried out in the forests in the interior of British Columbia indicated that the ribbed blades reduced damage to the butt ends of trees by approximately 80 percent compared to conventional smooth-sided shear blades (McLauchlan and Kusec, 1975).

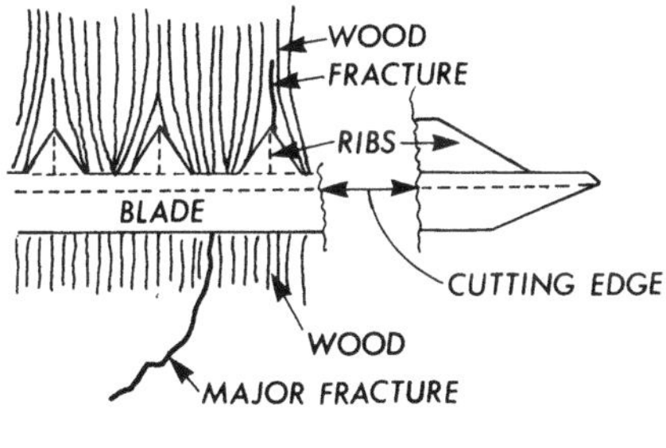

Figure 11. Shear blade with triangular ribs (McLauchlan and Kusec, 1975).

It was thought that it might be possible to reduce wood damage by inducing vibrations into the shear blades during the cutting process to reduce the friction between the blade sides and the wood and possibly concentrate more of the cutting force at the blade edge (Johnston and St. Laurent, 1975). The operating principle of the vibrator is shown in Figure 12a. The shears can be vibrated parallel or perpendicular to the direction of the cut. Tests of vibrating the shear parallel to the direction of the cut were conducted in Sweden by Wiklund (1967). The maximum cutting force, whether or not the blade was vibrated, was approximately the same. However the average cutting force required with a vibrated blade was considerably less. Vibrating the shear blade at right angles to the direction of shearing appeared to have no significant effect.

a)

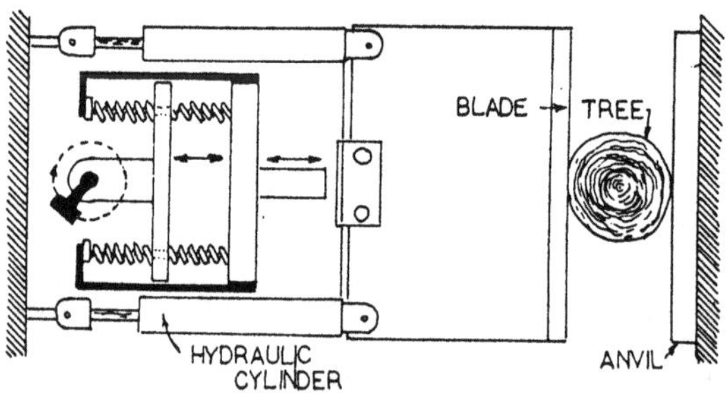

b)

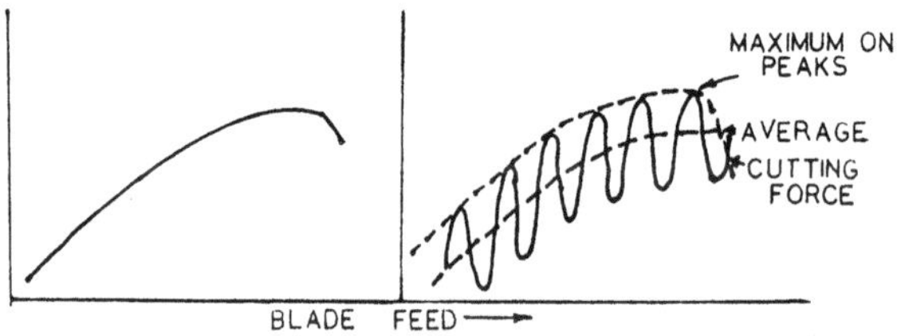

Figure 12. a) Operating principle for vibration shear blade. b) Cutting force without vibration and with vibration in feeding direction.

4.2.4 CIRCULAR SAW AND CUTTING DISC DEVICES. Because of the actual and perceived damage to the butt of trees when shear-felled, other mechanisms were sought. Wood damage was the dominant factor and not power requirements as it was found that the energy to fell a tree by chainsaw or by shears is approximately equal. In Scandinavia development lay mainly in the direction of hydraulically powered chainsaws, incorporated into the felling grappling head of a felling machine. In North America a number of mechanisms were developed involving circular saws and cutting discs incorporated in the boom-mounted grapple of a felling machine

(Folkema, 1984; Bjerkelund, 1983). One manufacturer mounted a specially designed 140 cm saw on their felling head with a side opening and operated it as a scythe. Trees are not grappled until the felling cut is complete, to minimize stress in the tree. During felling the tree is retained in an upright position on a butt support plate located just above the saw but behind the saw teeth. Another manufacturer mounted a 139 cm diameter saw in its felling head. The boom-mounted felling head is placed on the ground at the base of a tree and the entire head moves forward on a roller track to sever the tree from its stump. The head does not grapple the tree until it has been completely severed and the tree remains upright with its butt resting on a support plate above the saw. Because of its design and operation this tree feller can utilize regular circular saws instead of specially designed saws. Yet another variation employs two 61 cm saws side by side. This design lacks a support plate for the butt of the tree so the unit must grapple and hold the tree upright while it is being felled. Another design uses a felling head consisting of two saws stacked one above the other driven by a chain between them. A splitter or wedge is mounted in front of the arbor to permit it to pass through the tree being felled. This device has a sawkerf of 95 mm.

Besides tree fellers using saws there are a number which use cutting discs. The latter differ from saws in that the kerf is normally greater, up to 40 mm, but also because they operate with a high torque and at a much lower rpm, i.e., 150-300 rpm instead of 1200 rpm.

Because of the development of these various cutting devices the use of shears for felling trees has declined drastically. The felling heads are more complex than those with shears but designs are continually being improved and reliability is high.

4.3 CONCEPT AND PERFORMANCE

There are two dimensions to a logging system (whether one machine or more) which are critical to an analysis and understanding of any system in establishing its productivity and cost (Boyd and Novak, 1977). The first dimension is concept and the second is performance.

By concept is meant the functions performed, the arrangement of components to carry out such functions, the size, weight, cost, power required and fuel consumed according to the design developed to express the concept. For example, the concept may be to design a harvester which will produce tree lengths in uniformly sized piles within the forest stand. The concept may be to fell the tree mechanically, lower it under control into a horizontal position, remove its limbs and top and place the bole on the ground in a pile. An alternative concept might be to remove the limbs and top from the tree while it is still standing and then sever it from its stump and place the tree length on the ground in a pile.

Performance within the potential of the concept is the measure of engineering design skill and quality of manufacture which constitute the level of achievement in converting the concept into hardware. Performance will also include a measure of operating and maintenance skills and the suitability of application of the system. For example, having established the operating cycle times, the concurrent and consecutive operations, the design of the controls and the fabrication of the machine permit the operator to achieve the designed productivity.

Some characteristics of logging machines which reflect the concept from which the machine developed are:
- operating cycle time per tree
- machine size, weight and price
- power and fuel consumption. The functions to be performed dictate the power required, the duty cycle of the engine and these two in turn establish the fuel consumption.
- application factors. The machine concept sets the limits of mobility, travel speed and other design features such as tree size and species that can be harvested.

Concept factors or characteristics are limited to values that can be physically achieved. The values for such characteristics are not equal for all machines but they can only vary through a restricted range as they are determined primarily by the laws of science.

Performance factors are relatively unrelated to the functional ideas or concepts from which the machine developed. These factors include:
- engineering and design or the translation of the concept into hardware
- the quality and reliability of the machine components
- the operator and maintenance training and support facilities provided
- the application of the machine under the conditions for which it was designed
- all the human inputs at all levels which produce efficiency.

4.3.1 RELIABILITY. Reliability of machines and their components is critical to performance. It is important to remember that component reliability must be considerably higher than total reliability.

Reliability has been defined as "the probability of a device performing its purpose adequately for the period of time intended under the operating conditions encountered." Reliability can be expressed mathematically throughout the entire life of a component (Mellgren 1978).

When a component is operated within the period of its life which is called 'useful life' it will have a practically constant 'mean time between failures' (m). If it is an arbitrary 'operating time', then reliability (R) of a component may be calculated

$$R = e^{-\frac{t}{m}}$$

in which e = 2,718 (the base of the natural logarithm). For example the reliability of a component during an 8-hour shift (t) for different mean times between failure (m), in hours has been calculated as follows:

m =	8	80	160	200	230	400	800	8000 h
$R = e^{-\frac{8}{m}}$	0,37	0,90	0,95	0,96	0,97	0,98	0,99	0,999

After the "useful life" the above formula is not valid. The reliability of a complex system of components which forms one machine can be expressed as the product of the reliabilities of those components on whose satisfactory operation the machine depends.

If a machine has n components in series, the total reliability then is

$$R_{total} = R_1 \cdot R_2 \cdot R_3 \dots\dots\dots\dots R_n$$

in which R_1 = reliability of the first component in the machine system. In a series system machine if one component fails, the whole machine will fail. For example, assume a hypothetical harvesting machine with 10 component assemblies, with the following reliabilities (Axelsson, 1972).

$R_1 = 0{,}99$	$R_6 = 0{,}98$
$R_2 = 0{,}98$	$R_7 = 0{,}96$
$R_3 = 0{,}89$	$R_8 = 0{,}97$
$R_4 = 0{,}97$	$R_9 = 0{,}98$
$R_5 = 0{,}99$	$R_{10} = 0{,}99$

$$R_{total} = R_1 \cdot R_2 \cdot R_3 \dots\dots\dots\dots R_{10} = 0{,}817.$$

Reliability can be measured on a scale of 0 to 1. The assumption of a constant failure rate is the equivalent to an assumption that failures are exponentially distributed in time. This exponential failure distribution model is considered to be appropriate for advanced logging machines because systems are found to exhibit a failure rate that is essentially constant over the period of their operating life. For equipment operated in the time period of constant failure rate, the reliability of the equipment is expressed by the exponential relationship

$R = e^{-\lambda t}$ in which
R = reliability (probability of no failure for a time t)
t = specified time period
e = base of natural logarithms (2,718)
λ = failure rate

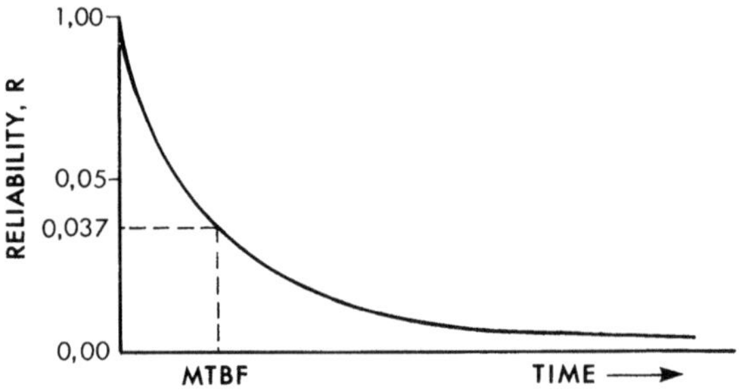

Figure 13. Reliability exponential relationship. The probability of an item surviving an operational time equal to its MTBF (mean time between failures) is 0,037.

Serious consideration should be given to the reliability of components and of machine systems when considering the introduction of single- or multi-function machines. The more complex a machine is and the more operations it is called upon to perform, the shorter will be the mean time between failures.

Table 5. Reliability of six harvesting machines (Mellgren, 1978)

Components	Feller	Delimber Slasher Sorter	Forwarder	Feller Delimber Slasher Sorter	Feller Delimber Slasher Forwarder	Flail Delimber
Frame	0,99	0,99	0,99	0,99	0,99	0,99
Engine	0,98	0,98	0,98	0,98	0,98	0,98
Drive Train	0,96	0,96	0,96	0,96	0,96	0,96
Feller Boom	0,98			0,98	0,98	
Feller Head	0,98			0,98	0,98	
Delimber		0,98		0,98	0,98	0,98
Slasher		0,99		0,99	0,99	
Sorter		0,98		0,98		
Stacker					0,98	
Hydraulics	0,97	0,96	0,98	0,96	0,95	0,98
Electrics	0,98	0,97	0,99	0,97	0,96	0,89
Loader Booms		0,98	0,98		0,98	
Loader Grapple			0,98	0,98	0,98	0,98
Total	0,85	0,80	0,87	0,80	0,75	0,89

Table 6. Estimated operational efficiency for tree harvesting - %

Machine	Excellent	Good	Fair	Poor	Very poor
Harvester	90	81	69	49	34
Feller forwarder	93	86	76	59	42
Limber bucker	91	83	71	51	36
Feller buncher	95	89	79	62	45
Forwarder	93	86	76	59	42
Landing delimber	92	86	75	58	41
Slasher	93	86	76	59	42

It has already been stated that there is no apparent conceptual advantage of importance between any of the machine systems under discussion. To account for the wide variations in productivity and cost, consideration must be given to the skills and efficiency of the operators and mechanics servicing these machines. The operational efficiency must be considered, along with the mechanical reliability to give an accurate picture of machine availability. Table 6 was developed by J. Kurelek (1976).

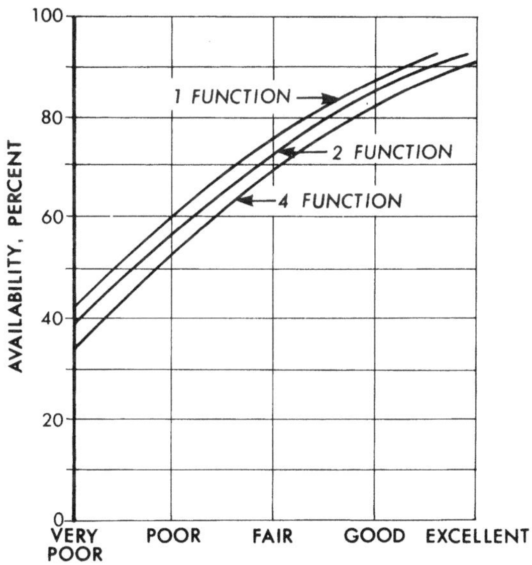

Figure 14. Machine availability vs operation competence (Kurelek, 1976).

Figure 14 shows the impact of operational efficiency for machines which perform varying numbers of functions. It can be seen that with a "fair" rated level of operational efficiency a single-function machine would have approximately 13 percent greater availability than a four-function machine. It also indicates that a single-function machine can tolerate a lower level of operational efficiency than can a multiple-function machine.

Besides machine availability, the efficiency of planning supervision of the harvesting operations must also be taken into account when estimating the productivity and cost of a machine system. The efficiency of an operation is usually calculated by dividing the planned or scheduled shift hours by the productive machine hours to give a figure that is representative of the shift output of the machine. Scheduled machine hours are those that a machine is scheduled to work. Planned non-productive times during a shift - personal, servicing and travelling - should be included in the scheduled operating hours. Machine productive hours are those hours of scheduled time that a machine actually works.

A machine may be scheduled for 8 hours operation but with a mechanical reliability of 0,85 it will actually only be operating for 0,85 x 8 = 6,8 hours. The efficiency of the operation, which involves planning, organization and supervision may only be 0,81 so that the 6,8 hours is further reduced to 5,8 productive machine hours within the scheduled 8-hour shift.

Kurelek (1976) in his analysis of various machine systems established that he could not count on productivity being closely related to machine cycle times. He found that production rates established in this manner did not appear on the operations over weeks, months, or years. He also stated that the more ambitious the machine design in trees per minute or kilometres per hour the higher was the percentage loss in productivity.

4.4 EQUIPMENT ACQUISITION THROUGH OWNERSHIP, LEASING OR RENTAL

In introducing this subject there is a basic assumption and that is, that the equipment in question is required by the organization considering the alternative means of getting the equipment onto the job.

4.4.1 OWNERSHIP. The most common approach to obtaining equipment is to purchase it outright, that is through ownership. Theoretically, this should result in the lowest cost. A leasing company must borrow money at the bank or from private sources and in most cases does not enjoy the low interest rate available to many large corporations. In actual practice, therefore, the leasing company becomes an agent for the corporation in negotiating a financial deal for its equipment. This is a generalization which will be enlarged upon later.

If equipment is purchased, the purchase may be made with money, the source of which may be equity capital, usually internally generated, or with debt capital.

The financing of equipment purchases with equity capital can be expensive. The cost of using internally generated capital for the purchase of equipment is the foregone earnings in the best investment opportunities of the same risk, either inside or outside the company. The condition for reinvesting company funds internally in this manner is that a plowback will add more to the investment earnings of the organization than other investments. This is strictly the economic approach because often equipment is acquired because it is essential to the operation of the business rather than because the investment in it resulted in a higher return than alternative opportunities.

New equity capital obtained through the sale of stock will probably be worth 10% or 12% or more. This would include the cost of floating the stock issue as well as the value of the shares.

Internally generated equity money used to purchase equipment requires that more than double the amount required must be earned, i.e. to purchase an $85 000 skidder with 54% corporation tax, some $157 000 must be earned to generate the necessary funds.

Debt financing can be a relatively low cost source of money compared to equity financing. It has the advantage that its cost may be deducted from ordinary income as an expense. However, the borrowing capacity of a company may be limited by its credit standing, or for other reasons. Debt of this nature appears in the balance sheet and adversely affects the rates of current debt to total equity and this, in turn, can affect a company's credit. This is one of the leading arguments of leasing companies.

The outright purchase of equipment is considered to have ample justification where certain factors are present. These are:

a) Adequate availability of funds from working capital requirements, equipment procurement, and other capital additions;
b) Ability to borrow money at reasonable rates of interest;
c) Optimum earnings on capital (borrowed or equity) invested in equipment;
d) Favourable tax results;
e) Favourable financial and operating ratios;
f) An efficient operation with the right equipment in the correct quantities for the job;
g) An efficient maintenance department (if lease agreement includes maintenance);

h) A minimum amount of downtime of equipment incurred;
i) Overhead costs including such items as executive time, clerical time, obsolescence, etc. fully controlled and in line with industrial standards.

In the case of small organizations, or large ones short on working capital, it may be necessary to pay for the equipment on the installment plan or rental purchase but no matter what the method, the intent is to obtain title to the equipment.

It is often argued that it is the use and not the ownership of assets which produces profits. This is obvious with the rise and development of impersonal corporate enterprise, the subjective prejudice of owning versus rental or leasing can be put on a more objective basis. Formerly when the entrepreneur rather than professional corporate management dominated the business scene, there was considerable status, both in business and socially, in ownership and clear title to assets that doesn't exist today. It was and usually still is considered to be the natural thing to purchase all the future services of a given asset or piece of equipment at once rather than buying such services as required.

It may be argued that if you purchase a large costly piece of equipment, for example, an $85 000 skidder, and pay cash for it, you are actually prepaying for the capabilities of this machine in its second, third and fourth years to the end of its useful life. This line of reasoning is used by leasing concerns when they propose that a piece of equipment should only be paid for as it is used and that capital should not be tied up in the form of prepayment.

4.4.2 LEASING. Equipment leasing in almost all fields of manufacturing is commonplace.

Some of the various pros and cons of leasing are:

(1) Leasing will generally provide 100% financing for the equipment to be acquired. This means that there need be little cash outlay at the beginning of a lease, usually just the regular periodic payment at the end of the first month. This is also generally true if equipment is acquired on the rental purchase basis where rentals are credited towards the purchase price of the equipment.

(2) Because there is little cash outlay, a lease will free or conserve working capital for more profitable alternative investments. Old equipment can be replaced and new equipment acquired in order to cut costs, streamline operations, and increase production by a lease with no use of current operating funds. This conservation of working capital for alternative investment can be a worthwhile argument. However, the control of cash and cash-flow in a company is of great importance. If the freeing of capital is the principal reason for leasing,

the working capital thus freed must be put to work. If it isn't, this advantage is lost. Too much cash in the company treasury is undesirable. Cash is usually the least productive asset a company has in terms of the yield it returns. Companies with excess cash normally lend it out on short-term notes, thirty-, sixty-, and ninety-day loans at rates currently of 6 to 8 percent per annum. This is a poor showing compared with a post-tax profit of perhaps 25 percent on current assets.

(3) It must be recognized at the outset that leasing is merely an alternative form of financing, comparable to and competitive with equity and debt financing. It is obvious that a lessor, under his lease agreement, will charge a sum sufficient to reimburse him for the difference between the cost of the asset leased and its residual or scrap value. Interest on the capital invested in the asset will be included plus a reasonable profit. To be on the safe side, the lessor may underestimate the residual value of the equipment or even recover the full cost of the equipment plus interest over the term of the lease. This along with the profit margin represents an increased cost to the lessee over any other method of acquiring the equipment. If, on a basis of direct comparison, leasing costs more, why lease?

(4) Leasing can be used to improve a company's financial statement. As mentioned earlier, debt financing appears in the balance sheet and adversely affects the ratio of current debt to total assets. To date it has not been necessary to indicate any place on a balance sheet that a company was involved in any long-term leasing with long-term debt and an insistence that leases be disclosed. It is becoming common now to indicate lease commitments as a footnote to the balance sheet.

(5) Leasing may permit deferment of taxes. However, there is no tax evasion through leasing. By leasing, tax payments are less in the earlier years, and this postponement of taxes may have value to the lessee in that money available from such postponement can be used to earn more money until taxes are finally due.

(6) There are several different kinds of lease agreement. It is possible to have a lease with an option to purchase. This is similar to a conditional sales or rental purchase contract because, unlike a true lease, title passes to the lessee for a nominal sum at the end of the lease period. This type of lease does not have the tax advantage mentioned above because the equipment is treated as a purchase from the date of the lease, must be capitalized and depreciated accordingly.

A lease without an option to purchase is considered to be a true lease. The title to the equipment never passes to the lessee during the life of the agreement. Usually at the end of the lease period the equipment goes back to the lessor for disposal. It is legal, however, for a lessee to take title to the equipment at the end of the contract as long as such action is not a part of the agreement.

A lease with guaranteed maintenance is fairly common in large urban centres for large fleets of trucks and cars. It is less common in the construction and logging industry but is often the reason that companies consider leasing at all. This may be considered service leasing as distinguished from finance leasing.

The only recommendation that can be made is to carefully assess each leasing proposal on its own merits, in the light of your company's financial position, mechanical maintenance capabilities, and the lessor's reputation.

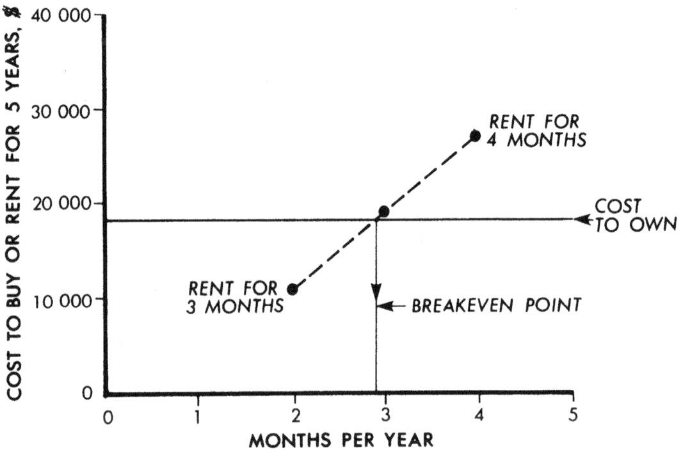

Figure 15. Cost of purchase or rent for 5 years at various annual usage. The chart is based upon a machine costing $70 000, a rental rate of 8% of purchase price per month, a trade-in value of the purchased machine of 10% and a machine life of 5 years.

4.4.3 RENTAL. From a legal point of view there is little difference between a leasing agreement and a rental agreement. However, leasing is considered to be for periods of 3-5 years while renting is considered for 1-3 months, a very short term.

The following are some typical conditions of rental of equipment.

(1) Monthly and daily rates are non-operated rates. Monthly rental begins when the rentee takes possession of equipment and ends when he returns possession to the rentor.

It is customary for the rentor to load the equipment on the rentee's transport.

(2) The monthly rate is based on an agreed-upon number of hours. After the first month, it is customary to prorate any additional period.

(3) The equipment should be inspected before being rented and be in good condition and be returned in the same condition as delivered, less normal wear and tear. Insurance on equipment rented by the month is to be carried by the rentee. On short-term rentals where excess cleaning and repairs are necessary, an additional charge should be made to cover such added expenses.

(4) The daily rental period begins when equipment leaves the rentor's yard and ends when it is returned to his yard.

(5) The hourly rate includes all repairs, cable, blades, tires, and running repairs, fuel, operator. Insurance is carried by the rentor on an hourly rate and daily rate.

(6) All rentals are payable in advance.

Short-term rentals can be an economical means of providing equipment for limited periods during the year to meet peaks in operating. A road construction job, extra haul trucks in winter, are examples. Short-term rental rates are high compared to calculated costs of ownership. Rates as a percentage of replacement value will run from 5-15 percent per month - trucks and tractors generally 6-7 percent, tractor bulldozers 8 percent, cranes 5 percent, light construction equipment 10-15 percent.

In spite of this high rental, and with the conditions of use as listed earlier, studies show that if annual usage is for short periods, up to 3 months/year with the equipment renting up to 8 percent/month, it is more economical to rent than to own (Figure 15).

It may be worthwhile for a company to make up its own table of rental rates versus ownership costs to establish for itself when it should rent and when it is justified in purchasing equipment.

4.5 DEPRECIATION

Depreciation is the diminution of value of equipment or other assets through use and other factors and contingencies. It is not applied to a resource which is subject to depletion in quantity. A carefully considered depreciation policy is necessary to reflect proper operating costs and to meet the demands of a business for the establishment of adequate reserves.

Depreciation may take one of a number of forms:

- a general wearing out or deterioration due to use or age, which cannot be made good by repairs, e.g. a house

- wear, tear and exposure to elements. This is a continuous process but the item seldom becomes 100 percent worn out. It may however depreciate to a point where it is not safe or economically useful. It is kept in serviceable condition through maintenance and repairs which often represent a fairly high percent of the original cost.

- sudden or accidental damage which may arise through accidents or unforeseen contingencies

- obsolescence is a form of depreciation due to changes which render the item uneconomical for use compared with more modern and more efficient ones. It operates regardless of wear, tear, accidents or other causes. It is often a major factor in depreciating certain pieces of equipment or capital improvements which are no longer required.

- abandonment resulting in the loss of useful value by the cessation of operations for which the capital improvements had been made

- deferred maintenance is the amount of the deferred expenditures necessary to keep the depreciable properties in the condition for ordinary efficient operations. Ordinary maintenance is not provided for as such in the annual charge for depreciation.

Depreciation is expressed by regular periodic charges against operating cost of amounts that fairly reflect the shrinkage in value of the capital assets involved, due to the exhausting of their usefulness or value arising from any of the above classes of lessening worth. Such a charge is a return of a part of the capital invested, to be continued throughout the predetermined beneficial life of the item, at the end of which the total of such charges, plus the salvage value, if any, should equal the original cost, plus any additions. Thus as the asset diminishes in value, the reserves increase by a proportionate amount. If the asset is retired from service due to accidental damage, before its normal life span, the reserve account will not be adequate to cover the cost of replacement but at least offers a partial provision.

Some organizations set up a sinking fund to have money for replacement available but more commonly the charges for depreciation are shown in the liability side of the balance sheet only.

Figure 16 shows a number of different ways in which to calculate depreciation. Each has its merits.

4.5.1 STRAIGHT LINE DEPRECIATION. Two variations of this method are considered. In the first no salvage value is considered, in the second a salvage value is taken into consideration. The advantage of the straight line method, where the capital cost of the item is divided by its life in years to arrive at a uniform annual depreciation charge, is based upon the perception that the difficulties of forecasting life and salvage value are sufficiently great that refinements in depreciation calculations are not justified.

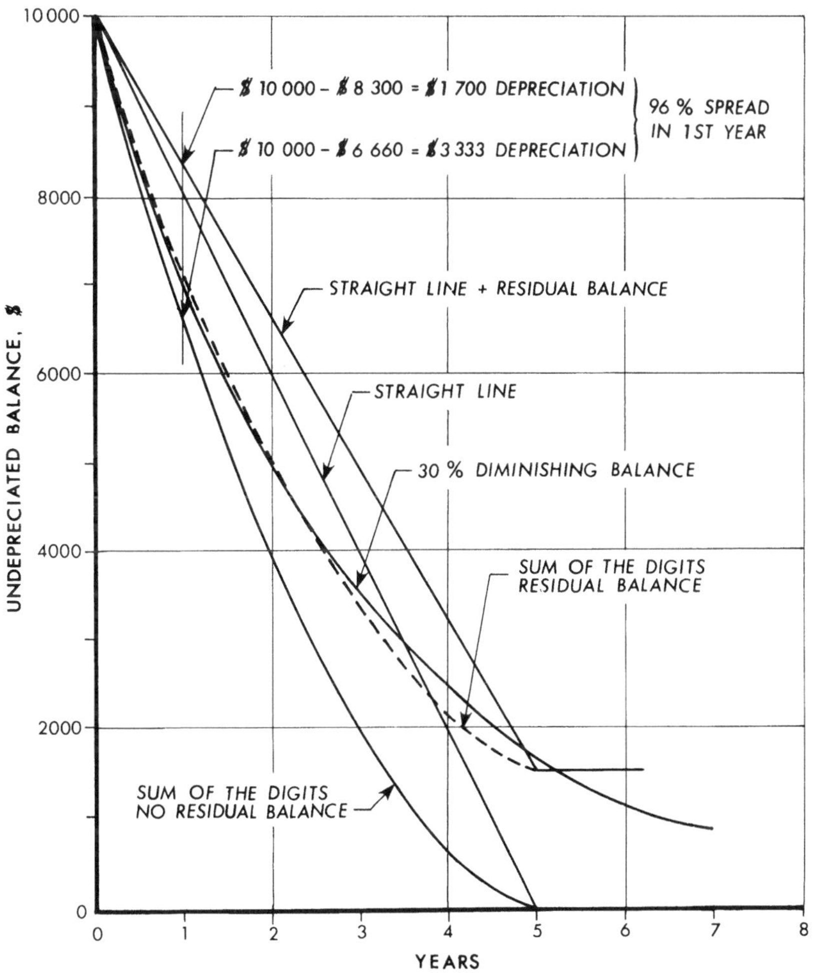

Figure 16. Various methods of calculating annual depreciation (Spanjer, 1968).

The establishment of a machine salvage value or trade-in value will depend in large degree on the physical condition of the machine, the conditions under which it has worked and the number of operating hours on it. Local conditions play a large part in establishing this figure as used equipment values vary widely around the world.

Whatever the variables, the decline in value is greater in the first year than in the second, and greater in the second than in the third and so on. The shorter the work life of the machine the higher the percentage of its value is lost in a year.

Although the percentage of loss each year depends on the life of the machine, as a general rule 40 to 50 percent of the machine's value will be lost in the first quarter of its life. By the halfway point of its lifetime, from 70 to 75 per cent of the machine's value will be lost (Caterpillar, 1976).

Table 7. Calculation of straight line depreciation - assume 5-year life. Purchase price $10 000; no residual value and $1 500 residual value

Year	Annual depreciation - no salvage value		Annual depreciation - $1 500 salvage value
	$		$
1	2 000		1 750
2	2 000		1 750
3	2 000		1 750
4	2 000		1 750
5	2 000		1 750
	$10 000		$8 500
		Salvage value	1 500
			$10 000

4.5.2 SUM OF DIGITS DEPRECIATION. Assume a machine life is 6 years. For each of these years assign a value beginning with 6 in the first year, 5 for the second, 4 for the third, etc. The sum of the 6 digits is 21. In the first year 6/21 or 24% of the initial purchase price will be depreciated, etc. In the sixth year 1/21 or 4% will be depreciated.

Table 8. Sum of digits depreciation. Assume a 6-year life, purchase price $10 000; no residual value and $1 500 residual value

End of year	Largest remaining digit	Loss in annual value in %	Annual depreciation in $	Remaining value in %	Remaining value in $
1	6	29	2 900	71	7 100
2	5	24	2 400	47	4 700
3	4	19	1 900	28	2 800
4	3	14	1 400	14	1 400
5	2	10	1 000	4	400
6	1	4	400	0	0
21		100%	$10 000		
		No residual value.			
1	6	29	2 465	71	6 035
2	5	24	2 040	47	3 995
3	4	19	1 613	28	2 380
4	3	14	1 190	14	1 190
5	2	10	850	4	340
6	1	4	340	0	0
21		100%	$ 8 498		
		Salvage value	1 500		
			$ 9 998		

4.5.3 DIMINISHING BALANCE DEPRECIATION. A common method of calculating annual depreciation is to reduce the value of a machine by 30 percent on the diminishing balance. This method results in the heaviest depreciation in the early years, when repairs and maintenance costs are lightest and are lightest towards the end of the machine's life when repairs will be heaviest. This tends to even out the total charges for depreciation and repairs.

In using this method the annual depreciation charges will never reach zero and it requires a salvage value greater than zero.

Table 9. Calculation of depreciation on a diminishing balance. Assume a 5-year life, purchase price \$10 000; no specific residual or salvage value; 30 percent annual depreciation on diminishing balance

Year	%	Annual depreciation Annual value remaining = \$	Annual amount depreciated = \$
		\$	\$
1	30	7 000	3 000
2	30	4 900	2 100
3	30	3 430	1 470
4	30	2 401	1 029
5	30	1 681	720
			\$8 319
		Salvage value	1 681
			\$10 000

The \$1 681 shown as a residual value is referred to as the book value of the asset.

4.6 EQUIPMENT REPLACEMENT

The replacement of harvesting equipment, i.e. saws, skidders, loaders, is essentially a problem of the control of internal capital investment. Decisions on equipment replacement are normally a part of the capital expenditure budget designed to ration limited capital funds among alternate investment projects. New equipment must compete for funds with alternative capital projects and must prove its acceptance on the basis of a rate of return in investment (R.O.I.) that is superior to alternative uses.

In large organizations the estimated cost saving over the old equipment is not considered to be an economic justification for replacement. The productivity and cost of the output of the new equipment must be enough cheaper that the prospective cost savings earn a higher return on the invested capital than can be obtained on the marginal investment elsewhere in the organization. For a small operator where there is little capital and few items to spend it on, the above principle still applies, but cost savings of new over old equipment may be sufficient to justify replacement.

When considering equipment replacement, the total applicable cost to keep the old unit for one more year is compared with the average yearly cost of the new replacement unit over its chosen lifetime. The average yearly cost of the new replacement is used rather than just the first year cost because once a decision is made to purchase a new unit a commitment is made to its costs for its chosen lifetime. If the first year costs only were used the higher costs incurred towards the end of the unit's life would be ignored. To retain an old unit is a year-to-year decision in which past capital spending or repair costs do not apply.

Past records do not enter into the analysis as to whether or not to replace equipment except as an aid to evaluate and estimate future costs. Sunk costs, both in terms of capital as well as earlier repair and maintenance expenditures have no place in the projection of future costs. The original acquisition cost of a machine that is now in operation is sunk and unrecoverable. Book value costs are of value for bookkeeping or income tax purposes. No matter what the machine's original cost was, when it was purchased, and how much depreciation has been charged off on the books, the only real depreciation cost that is important is the future decline in the disposal value of the unit over the future period considered in retaining a unit in operation.

4.6.1 REPAIR COSTS. In the case of the old unit repairs would include whatever overhaul and running repair costs are necessary to put the machine in such condition that there is a reasonable expectation that it will be operable for one more year.

Repair costs for the new unit should generally be obtained from past records to comply with its apparent lifetime. Past costs should be brought up to present day prices and then be projected into the future to find the estimated future repair cost of the new unit. Where possible the average costs for classes of units rather than those for individual units should be used to determine the new unit's future costs.

4.6.2 UNAVAILABILITY. Unless a machine is in very poor condition it is normally difficult to justify its replacement on the basis of repair costs alone. There are other factors of importance that must be considered in such an analysis. Unavailability and productivity of a machine are legitimate cost factors, even though they are difficult to estimate.

As a machine gets older it may be assumed that unavailability increases. Unavailability depends upon the general state of repairs of the old unit. As a result of its poor condition extra units may be held in reserve in case the unit breaks down, the remaining units may be additionally loaded to make up for the progressive loss of production by the old unit, extra hours may be worked by the remaining units to make up for loss of production.

To place a monetary value on a unit's unavailability, if a replacement is rented, it is the difference between the costs of the rented machine and the old unit plus those fixed machine costs that continue, whether or not the old unit is working. If a standby machine is included in a fleet or group of machines then the fixed costs of this standby machine should be prorated among the working units. If it is not working there will be no operating costs. The shared portion of the fixed cost of a standby machine will be a conservative figure because if the standby unit was working it would be earning a profit and this loss of profit should be a cost of unavailability.

If a machine is down and unavailable and daily production on the job is reduced, the fixed overheads are distributed over fewer items of production and unit costs are increased. It may be possible to establish how much longer operations have to be kept going as a result of the cumulative effects of production loss and this cost should be charged against machine unavailability.

If estimates of cost for unavailability are to be developed it is essential that records be kept to give a basis for such estimates. Judging the general condition of an old unit, coupled with past records of unavailability should make it possible to estimate the unavailability cost for the next year.

Similarly the unavailability costs of the proposed new unit over its lifetime must be included. Past records would be the source of such information. It is treated and applied in the same way as repair costs of the new unit, where total unavailability costs would be added to total repair costs before averaging. As a machine ages, the cost of unavailability can approach the cost of repairs.

4.6.3 PRODUCTIVITY. In many instances productivity of a new unit is higher than that of the old. This is particularly true of tractors and skidders where engine power continues to rise almost year by year. In such cases the old unit is penalized for its inability to do the same work and have the same output as the new. If actual work measurement cannot be established in the field it is suggested that the engine power ratio of the two, multiplied by the standard machine hourly operating rates be used to arrive at the old unit's operating deficiency. The difference in productivity is charged against the old machine.

4.6.4 FUEL. Fuel costs enter into the analysis through a difference in kilowatt ratings or where diesel oil may be used instead of gasoline. In every instance the old unit is charged or credited, as the case may be. In the event that the kilowatt rating of the new unit is higher, the old unit could be credited with the cost of the fuel.

4.6.5 LOSS IN SALVAGE VALUE. If an old unit is kept for one more year there will be a loss in salvage value. The older the machine is the less significant its salvage value becomes. The loss in salvage value may be offset to some degree through a depreciation credit by keeping the old unit for an additional year. If depreciation is applied on a 30 percent diminishing balance it means that the first year depreciation of 30 percent of the original cost is declared an expense, 21 percent the second year, 14,7 percent the third year, 10,3 percent the fourth year, 7,2 percent the fifth year, 5,0 percent the sixth year and so on.

As an example assume that the difference in salvage for a 5-year-old machine is 4 percent between the fifth and sixth years. The difference in depreciation between the fifth and sixth years would be 7,2 - 5 percent or 2,2 percent. The old machine would be debited with a salvage loss of 4 percent and credited with depreciation of 2,2 percent for a net loss of 1,8 percent converted into money values.

4.7 RELIFING FORESTRY MACHINES

Relifing of existing equipment is not a new concept but it is receiving more attention as the price of new machines escalates. The application of this concept to machines used in the forest is an alternative to replacement by new machines. This alternative is receiving greater attention due to the high cost of equipment replacement. In the United States, it has been shown (Cubbage and Werblow, 1985) that the purchase price of logging machinery has increased fourfold in the past 18 years. Tufts et al (1981) found that the general price rise for harvesting equipment closely followed the inflation rate of the economy.

Table 10 shows some average equipment purchase prices by category and by year (Cubbage and Werblow, 1985).

As has been stated earlier when a machine or machine component is operated within the period of its "useful life" it will have a practically constant mean time between failures. After the useful life this does no longer apply. Initially, problems are usually encountered in the introduction of new equipment but once these are resolved the equipment tends to operate with a practically constant rate of maintenance until its useful life is over. Figure 17 (Axelsson 1972) shows this clearly.

Relifing involves the reworking or replacement, as appropriate, of all the functional and structural components of a piece of equipment in order to restore it so that its performance characteristics and remaining life expectancy will equal that of a new item of the same type (Eryon, 1981).

The relifing process involves the disassembly of a machine, inspection of its components, the rebuilding or replacement of components where indicated, reassembly and testing.

For forestry equipment it has been suggested that the cost of relifing should run between 40-60 percent of the cost of replacement with a new machine. Using data presented in Table 10 if a cable skidder with a 5-year useful life is to be replaced in 1984, it will have been purchased in 1979 for \$46 000. The current price of a new machine of comparable specifications would be \$62 000 while a proposed relifing would cost say \$62 000 · 0.60 = \$37 200. This is less than the original cost.

Table 10. Average equipment purchase price for various years

Equipment	1967	1972	1974	1976	1979	1980	1982	1984
				Years $ U.S. ($000)				
Chainsaw	0,20	0,26	0,30	0,34	0,44	0,46	0,42	0,55
Feller Buncher -tracked	-	40	75	90	120	120	-	188
Cable Skidder 86-115 HP	-	19	23	29,2	46	41	56	62
Grapple Skidder 101-129 HP	21	29	34	54	62	62	79	89
Farm Tractor	5,5	6,5	9	-	27	-	-	36
Forwarder 80-100 HP	-	18	28	38,5	40	44	45	63,5
Loader - Knuckleboom	10	12,5	16	20	24	27	29	27,5
Whole Tree Chipper 300-400 HP	-	-	-	-	82	82	112	137
Diesel Truck - Tractor	18	24	26	41	44	48	53	70
Shortwood Trailer	4	3,8	5	6,5	7,6	8,5	10,4	11

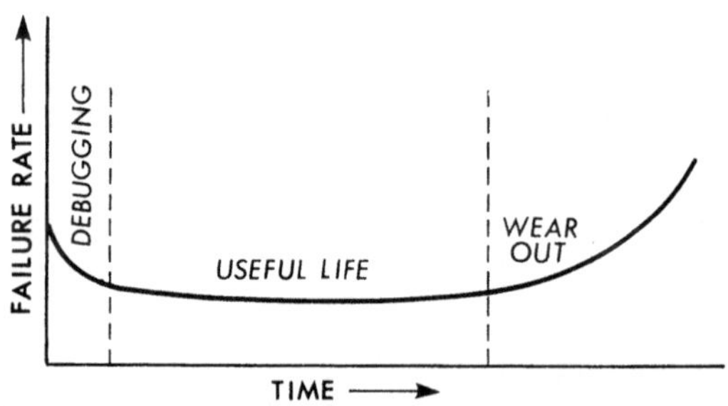

Figure 17. The "bathtub" curve. Normal development of the failure rate over machine life. The failure rate is approximately constant during the period of useful life.

From an accounting point of view relifing a piece of equipment is of interest as the cost of such an operation can be costed into the current year's budget rather than having to be capitalized as is the case with the purchase of a new piece of capital equipment.

The decision as to whether to relife equipment requires a detailed examination of the existing and potential new equipment and factors such as the following should be taken into account:

- capability of maintenance staff and facilities
- presence of extensive structural fatigue and corrosion
- downtime of existing equipment during relifing versus delivery time of replacement
- relative merits of carrying relifing in-house versus contracting out
- consideration of improved productivity or reduced operating costs with replacement
- value of existing repair parts inventory.

5. CONTROL IN WOODLANDS OPERATIONS

5.1 CONTROL FUNCTION

Woodlands operations, specifically logging, are facing an increasing need for control. As the cost of logging machinery skyrockets and labour costs increase, the need for tighter controls on operations increases in direct proportion. Many operators are instinctively cost and production conscious, but many do not understand the value of control on their operations. The black book in the hip pocket type of control is a thing of the past and, if an operator is to survive, he must tightly control his men and machines to ensure their maximum productivity and efficiency. One definition of business is that it is the survival of the least inefficient.

With reference to the sawmill and pulp and paper industry, it is suggested that the largest profit variable for most companies will be wood policy. Today, wood amounts to 50 percent of the cost of raw material making up a tonne of pulp. It is proposed that "what is at stake are earnings swings of up to 25 percent, pulp mill capacity limitations, soaring wood costs. To deal effectively with such a situation will require a major change in traditional woodlands management approaches of most firms, in particular a longer-range view, much more use of scientific techniques of inventory control, forecasting, site analysis, transportation, etc." (Koplow, 1974).

The word control has two meanings both of which apply to control as discussed herein. Control is the act of controlling, restraining or directing influence; it means regulation. It also means a standard of comparison against which to check results of action. From this then, control in logging requires the development of standards against which to measure results of action and following upon such comparison to regulate future actions in relation to the standards.

Control of a production process within a pulp and paper mill has been defined as the ability to bring the process to any given state and to maintain it there (I.B.M. 1961).

The purpose of control may vary in timber harvesting. It may be to increase productivity, improve quality, reduce costs, to obtain the fullest utilization of the equipment available, or any combination of these.

In many processes, control is now automatic in that an instrument records a property (heat) and when in the process, the property (heat) exceeds a given limit the instrument reacts upon the heat source to reduce the property (heat) to an acceptable level. A thermostat in an automobile or in a home are every day examples of such control.

Control in woodland operations is managerial control and is primarily a function of administration. Although managerial control cannot be affected as rapidly as process control, the principle is identical to that in process control. In managerial control, the data collection system gathers data, determines which are within specified limits and which require correction, and acts to regulate the latter (Den Uyl, 1968).

The industrial engineering approach to operations flourishes best in industry with a highly developed scientific outlook and timber harvesting until very recently was far removed from this attitude.

Woodlands departments of companies in general have been given little management attention by executives of pulp and paper companies in the past (Gauthier, 1968). This situation resulted from many factors, the most important being:

a) the location of woodlands activities in remote areas of little interest to head office personnel;
b) the relatively low capital investment traditionally associated with woods operations;
c) the difficulty for individuals associated mostly with office or mill work to understand the witchcraft of harvesting operations.

The result of this has been that woodlands organizations have lagged in the use of modern management techniques which are common in other areas of the same enterprise.

Where control existed, it was usually carried out by rule of thumb. Early attempts to introduce control into woodlands operations were pioneering efforts to introduce industrial engineering concepts and improved planning into operations which heretofore had been relatively slackly administered. The principles of control introduced were the same as those operative today. Control at that time was defined as an organization working in close co-operation with all departments toward the conservation of resources, improved logging practices and reduction of wood cost through principles of effective analysis, investigation and service.

Labour, wages and provision costs were low and timber in the early days of logging was plentiful and easily accessible. As long as there was a safe margin between logging and milling costs and current sales prices, work proceeded on its way and little time and consideration were given to establish the probable cost of operations. In the industry as a whole, the managements were not acquainted with the actual logging costs of individual areas nor with the reasons and factors which contributed to such costs.

Initially production standards were based upon average production under average local conditions. These average conditions included average climatic, forest and labour conditions. Today the use of averages which include built-in inefficiencies is not acceptable. Rather the approach is to synthesize an operation from its components, making necessary allowances in any specific operation for variations due to labour, climate, equipment, efficiency of the operation, etc.

The value of control is indisputable as has been proven wherever it has been applied. It is possible to introduce a system of tight control on productivity and cost on woodlands operations. Such control is essential under all conditions, when prices are high and production rather than cost may be the dominant factor, or when prices are depressed and cost becomes all important.

Control in logging is essentially the common sense application of reliable cost and production data to day-to-day logging problems. At lower organizational levels the sphere of decision making or the number of optional courses open are relatively limited. They become more complicated at higher levels due to the fact that there are a greater number of options possible rather than to the complexity of each.

At the top management level, controls may be exercised through the manipulation of funds; at the operating management level, control may be concerned with deciding how wood is to be produced or areas regenerated and how existing facilities (men, equipment, money) may best be used, and at the foreman level, where cutting or regeneration area, equipment and forest labour are all assigned, control may be largely a matter of volume (Harkness, 1963).

It must be established at the outset that control is not accounting, at least in the conventional sense. Accounting is history. It is the recording and distribution of past expenditures for product costing, for inventory control and for the preparation of year-end financial statements. It is highly regulated by "generally accepted principles" and is usually quite inflexible.

Control, including cost control, deals with a much broader area than accounting and is involved with improved operations, future planning, and conservation of resources. Production data are essential and normally the relationship between inputs and outputs are shown in purely physical terms, in contrast to cost and price data which show economic relationships only.

A form of control called responsibility accounting has been adopted by some companies. This form of accounting is not restricted by the "generally accepted principles" which are so important to financial accounting (Rouse, 1968). Responsibility accounting is a management tool designed principally for control purposes. Under this system, all accounting and statistical information concerning an organization's activities are classified according to the individual manager responsible for them, without undue consideration of product costing or financial reporting.

Responsibility accounting is based upon two precepts - definite responsibility for costs can and must be established and, second, a manager must only be responsible for the costs he can control.

The literature on control in woodlands operations is sparse. As has already been suggested, the industrial engineering approach to operations flourishes best in industries with a highly developed scientific output. The earliest and best known text on control is that by D. M. Matthews "Cost Control in the Logging Industry", American Forestry Series, McGraw Hill, Ltd. It was published in 1942, and has been out of print for about 45 years. Matthews' book was not used to any great degree on actual operations by forest engineering graduates. If it had been, the acceptance and practice of control in harvesting operations would be commonplace today.

The second book of consequence on this subject was "Planning and Control of Logging Operations" by Louis-Jean Lussier, University Laval, Quebec, 1961. This volume was based to a degree on the foundation prepared by Matthews but used more modern techniques of analysis. This volume went through several printings and it too is now out of print and unavailable.

The remaining literature on this subject consists of isolated papers and theses dealing with various aspects of control in woodland operations.

The above causes one to wonder if there is no interest in control as applied to woodlands operations throughout the industry or if other engineering texts dealing with control in general are sufficient and are being used.

If it is assumed that control is only suitable and practical in large companies and/or large operations with a highly trained technical staff, this is not the case. Control can also be usefully and profitably applied by small operators who must run an efficient business to exist in times when costs are up and prices down.

5.2 PRINCIPLES GOVERNING THE CONTROL FUNCTION

There are several underlying principles that apply to control as defined (Urwick, 1947). These are:

5.2.1 PRINCIPLE OF UNIFORMITY. All reports and data used for purposes of control must be in terms corresponding to the operation that the reporter can influence. For any given operation a person's authority and responsibility must correspond. In other words there is no place in control at a given level for the collection and reporting of data over which the reporter has no control, i.e. a sub-foreman need know nothing about current wage levels, or fixed charges on stumpage, etc., if such data are not affecting the operations for which he is responsible, or if they affect the operations he is in no position to modify.

5.2.2 PRINCIPLE OF COMPARISON. All reports and data used for purposes of control must be in terms of standards of performance. There must be standards or targets to be aimed at and any deviations from the target should be explainable. This is the feedback which indicates if any change in existing procedures is required and, if so, to what extent. The feedback may indicate for example that an additional haul truck is required, that a log loader has insufficient capacity, etc.

5.2.3 PRINCIPLE OF UTILITY. All reports and data used for purposes of control must be produced in a time period which permits their use by management. This must be compared with or be analogous to "real time" in a computer which is carrying out a control function in a process (Figure 18).

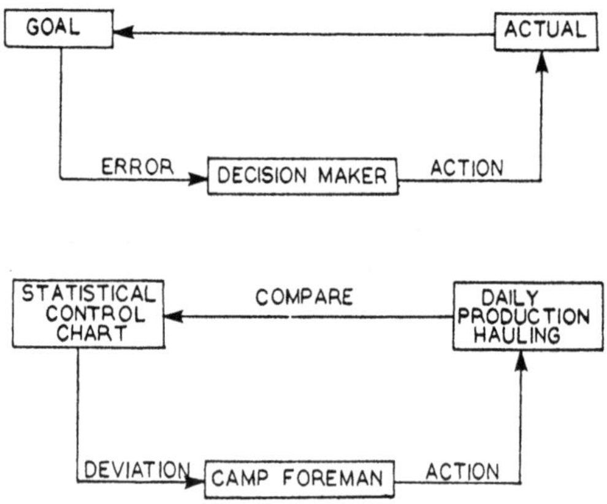

Figure 18. A flowchart of a self-controlling communication system. Actual performance is compared with goal. If it does not reach desired performance, a control mechanism acts to correct the actual performance until it meets the required standard.

It can be seen from the above that weekly or monthly production reports are not in a time frame that permits remedial action on, say, a yarding operation. Even a daily report may be too long a time frame. In some instances, a foreman should be able to tell by midday whether or not a crew is going to make an acceptable day's production. In this case, he can act immediately to rectify what may be a deteriorating situation.

5.2.4 PRINCIPLE OF EXCEPTION. All reports and data should emphasize deviations from the norm as it is these deviations which require consideration and action. As described earlier this is analogous to process control and is shown in Figure 19. A record of this nature permits a foreman or supervisor to establish the status of each day's production. Instead of drowning a supervisor in reports, exception reporting provides only relevant information. Exception reporting is analogous to process control. In process control, data are received and processed but no action is taken as long as information received remains within specified limits. It is only when data exceed the specified limits, is action taken.

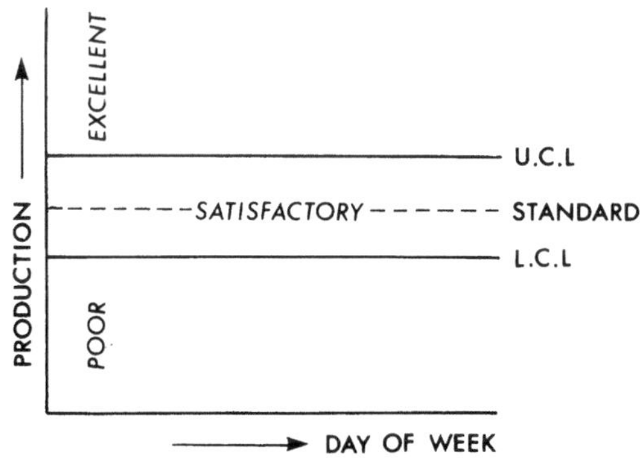

Figure 19. Production Control Chart, showing upper and lower control levels.

5.3 SYSTEMS APPROACH OR OUTLOOK

The growing, harvesting and delivery of wood in various forms for various end-products involve many operations which are distinct in time and place. The whole is a complex productive system which includes both inputs and outputs (Koten, 1969). The input to such a system is the output of a separate biological system (forest) which in turn is subject to variations in location, size, density and quality. The harvesting and delivery system is a complex production system composed of the output of the biological system, men, and machines and methods. Money could be included separately here but is actually included in men (wages) and machines (capital).

The forest harvesting system may be considered to be more complex than most production systems because of the necessity of moving the production machines to the input and hence encountering the various conditions imposed by soil, topography and climate associated with the site. Similarly manpower associated with a production or harvesting system will also be required to operate under constraints imposed by variation in the site conditions.

Harvesting systems are considered to be totally mechanized and systems analysis is concerned with the problems of compatability which results when many parts are assembled together to form a whole.

A harvesting system may be a multi-function machine made up of a number of different assemblies, the functioning of each being dependent upon the others; or a harvesting system may consist of a number of individual components or operations related to one another so that the operation of one directly affects the operation of the others. An example of the latter is a yarding operation where the log landing is of a size that, to permit continuity to the operation, it must be cleared and the capacity of the loader and the number of haul trucks and their scheduling are all integrated to permit optimum operation of the side.

Systems analysis permits the analysis of the probable outcome of each alternative course of action which will eliminate sub-optimization and permit one activity to be examined in the context of the whole operation.

At the present level of mechanization in forest operations, few pure systems exist as such, but systems analysis and the total cost concept are becoming widely accepted and are spreading rapidly and are aiding greatly in the acceptance of total mechanization. Total mechanization means an operation in which trees are felled and hauled out to roadside untouched by human hand.

5.3.1 HARVESTING SYSTEMS. It is possible to distinguish between a harvesting method and a harvesting system. A harvesting system has been described as consisting of individual components or operations which are so directly related to one another that the operation of one affects the operation of the others. One normally relates to the manner in which logs or bolts or other forms of product are produced. It may or may not be labour intensive or be partly mechanized, but is not normally considered to be completely mechanized.

It is inherent in a system that the various machine elements or machines are interdependent in action and the operation of one affects the operation of all the others. The efficiency of a system is, therefore, dependent upon the efficiency of each of the system components and is the product of the efficiencies of each. If a logging system consists of three machines, each of which has a mechanical reliability of 80 percent, then the overall reliability of the system will be 80 x 80 x 80 or 50 percent. If a multi-purpose harvesting machine has eight assemblies (loading boom with grapple, delimber, infeed rolls, debarker, outfeed rolls, shear, off-loader) each of which has a mechanical availability of 95 percent, the overall availability of the machine will be only 66 percent.

The availability of machine and components in a system, particularly for equipment operated in a forest environment, is a considerable problem.

Machine reliability can be established up to a point beyond which the engineering design and production are too costly. The cost of lesser reliability is less costly to live with than is the increasing capital cost of the more reliable unit. There must be a trade-off here.

As logging moves in techniques towards a specialized mass production system the workflow becomes more and more rigid.

The production line becomes clearly defined, whether in one multi-function machine or in a number of single-function machines. Buffer stocks are reduced and the length of possible waiting time between functions is reduced and in some instances is eliminated. Sources of input diminish in number to the point where in new systems one machine alone will provide the input for the next machine and so on. It is more and more difficult to change the workflow in the instance of a breakdown. In many instances, such a breakdown can paralyze the subsequent phases of the production system.

Table 11 shows the relative rigidity of workflow under different levels of mechanization.

At present, there are four different major systems of timber harvesting in various stages of development. Within each of these systems, there is a great variety in the components which make up each system. The timber harvesting systems are shortwood, tree length, full tree and chipping in the forest. The system titles describe the form in which the tree, once felled, is moved from the stump area to the roadside or landing. With regard to chipping in the forest, at present the chipping operation normally takes place at roadside or landing, but machines already exist which will chip trees at the stump.

The systems are described because their elements vary widely in sequence and timing and as a result the control data required will vary with each.

In the four different systems listed, the physical distribution of the operations to be performed vary widely. The shortwood system has the maximum number of operations carried out in the forest, the full-tree system has the minimum, while tree length and chipping in the forest are intermediate in character. These are illustrated in Figure 1 and have been described in Chapter 4.1.

In the shortwood system, tree sections in log (4 m plus) or bolt (3 m long) form are produced in the forest stand. In bolt form, the sections may be piled or bunched in the vicinity of the stump from which the tree has been severed. In log form the tree may be sectioned with the logs lying where the tree fell. At the present time in pulpwood size stands of timber on reasonably level ground the production of bolts has been successfully mechanized. The production of logs is still for the most part carried out manually with powered hand tools (chainsaws) and is highly labour intensive, particularly in steep terrain. There are distinct limitations on man-day productivity under such conditions. As labour rates increase, because of the high labour content, the cost of wood produced by this method increases directly. At present, there does not appear to be any possibility for economically mechanizing the falling and bucking of very large

Table 11. Level of workflow rigidity (Legendre, 1973)

Indicator	Traditional system	Semi-mechanized system	Fully mechanized system
No waiting time possible (versus yes)	No	No	To some extent
Single-purpose equipment (versus multi-purpose equipment)	No	Yes	Yes
Production or service line (versus no set line)	No	Yes	Yes
No buffer stocks and no delays possible (versus yes)	No	No	To some extent
Single-source input (versus multi)	No	No	Yes
No re-routing of work possible (versus yes)	No	No	Yes
Breakdown stops all workflow immediately (versus not all workflow)	No	No	To some extent
Breakdown stops some or all workflow immediately (versus no workflow stops)	No	Yes	Yes

trees into log lengths in steep terrain. There would appear to be more potential in bringing out tree lengths or full trees and processing them at the roadside or landing rather than in situ. In the manual production of bolts in the stump area, only a fraction of the working day is spent in essential operations such as felling, limbing, topping and cutting to length. Most of the day is spent in clearing trails, clearing pile sites, moving bolts to pile sites and putting up in piles or bunches. These latter operations actually add little to the physical production of the bolts, and, if eliminated, would result in increased man-day productivity.

In the tree-length method trees are felled, limbed and topped within the forest stand and are then skidded in this form to the roadside or landing for conversion into shorter lengths or to be further transported in tree-length forms. Normally there is no attempt to manually move or assemble tree lengths in the forest, although this operation may be carried out mechanically. In both instances of manual and mechanical operations, the productivity of the output will be greater for this system than for shortwood because fewer operations are performed upon the tree and wood is handled in larger units.

In the full-tree system, trees are felled and with no further work expended on them, they are draggged or carried out to roadside or landing for conversion there, or for transportation to a central processing plant. The concept of full-tree harvesting is not new, but the transformation of this method into a system is dependent upon the development of suitable equipment.

In the chipping in the woods system, trees are converted into chips either within the stand in the vicinity of the tree stump or at roadside or landing. Development of this system has been relatively slow and present limitations are high power requirements and relatively low productivity. Tree size and mobility of the chipping unit are probably the controlling factors in the success of this system. As is apparent, this system has application only where the end result is pulp or possible composite wood products. This system has a great potential dependent upon several factors not related to the harvesting system but to the end use of the product. As bark acceptance standards become freer and as techniques for the separation and segregation of bark from wood in chip form improve, the chipping of whole trees, limbs, tops and all will become possible. Yields per acre will increase greatly and the economics of this system will improve.

The number of machines and men making up each system will vary. It is obvious that the fewer machines and men involved the higher man-day productivity is likely to be.

There are many hybrid or modified harvesting systems in addition to these four major systems. Especially in regions with a dominance of farm forestry with small scale operations and resident forest owners, techniques are often chosen which meet particular owner objectives. These can vary widely depending on such factors as provision of employment opportunities for the owner and his machines, influences on his tax situation, environmental concern and so on. In some countries even legislation banning highly mechanized harvesting systems is considered in order to try to stop the depopulation of remote forest regions. All these adaptions can be lumped together in what may be called appropriate techniques for small-scale or labour-intensive forestry with additional objectives than those valid for large forest industrial enterprises. Considerable research is carried out in many countries to promote the development of such techniques.

5.4 WOOD INVENTORY CONTROL

Unlike many industries the forest industry carries much, if not most, of its inventory in the form of raw material, as logs, bolts or chips. To properly control wood inventory it is necessary to hold minimum safe inventories at all points of delivery which will be sufficient to ensure continuity of production at the shipping or conversion point. Inventory should be held at the stage which involves least cost and it should be possible to establish when excessive or too small inventories develop.

When wood is cut and piled in the forest it may be considered as undelivered inventory until it shows up in one of the subsequent stages of delivery. When delivered to the next point it ceases to be undelivered wood and becomes a more costly form of inventory. Regardless of whether the wood is moved by land or water it ultimately arrives at a mill where it is received and stored until it is used. Figure 20 is an example of such a wood flow.

FOREST (Standing tree)
↓
WOOD HARVEST
↓
UNDELIVERED INVENTORY (Wood in forest)
↓
STREAM INVENTORY (If water transport)
↓
MILL STORAGE INVENTORY (Sawlog deck or pulpmill blockpile)
↓
MILL CONSUMPTION (Lumber or pulp production)
↓
MARKETING OF PRODUCT
↓
UTILIZATION OF PRODUCT
↓
DISPOSAL
↙ ↘
WASTE SALVAGE

Figure 20. Flow process from raw material to end product.

If there was no trend in mill consumption and no seasonal variation caused by climatic or other physical factors then the time path from standing tree to mill consumption would be a straight line. Without any instabilities there would be no variation. However there are physical constraints on any wood flow process, usually caused by seasonal or climatic conditions. Seasonal constraints tend to be relatively simple and can be superimposed on a seasonally adjusted model.

Average seasonal variations month by month for past periods can be measured. They can be divided out of the actual data to leave as a residue the main non-seasonal movements inherent in the data which are cyclical in nature. When the dominant seasonal variations are removed the remaining non-seasonal movements are relatively smooth cycles. Actually there are normally two seasonal variations, that which is related to end-product needs and that in wood production due to physical constraints or climate.

With proper inventory and delivery controls it should be possible to permit the holding of minimum safe inventories at each stage that will ensure continuing operation at the next stage. It should be possible to reduce the most costly inventory at the shipping point or mill at the expense of earlier and less costly inventory farther back in the system.

Many forest industrial enterprises purchase a considerable share of their annual requirements of wood from other forest owners, as stumpage or deliveries at roadside, wood yard or mill. This supply is often less predictable than the flow of timber from their own forests. It often happens that a majority of sellers tend to overrun the contracted deliveries when price reductions are expected and shrink deliveries when price increases may develop. The control of the flow and inventories is then more difficult but also more important to safeguard against interruptions in mill operations because of a shortage of wood or against unnecessary high inventories of cut wood. In such situations the inventory control should also include assessments of external influences bearing on the attitudes and motivation of the sellers to fulfill their delivery contracts. Often buyers will refrain from enforcing contracts as they want to stay on good bargaining terms with sellers for the future supply of wood.

Also another phenomenon related to inventory control should be observed. Forest products are global commodities with considerable fluctuations in demands, generally related to overall economnic trends. There are many agents in the production and marketing chain from stump to end use. When the agents marketing for end use expect a decrease of demand they reduce their inventories and also their orders from those agents from which they buy. These agents in turn react usually in the same way and the reaction to signals from the end-use market will be transmitted all the way back to the forest resource. If all the agents in this chain react in the same way it is easily understood that there will be a type of multiplication effect on the market for raw wood material. Consequently, even minor fluctuations in demand of end users as a rule are amplified backwards. This phenomenon requires special skill and flexibility by the managers of wood flows and inventories. It also explains the often exaggerated fluctuation of wood and stumpage prices which often are not justified. This market mechanism can easily be observed for petroleum.

6. INFLUENCE OF THE FOREST ON OPERATIONS

To determine the degree of mechanization best suited for a region requires consideration of many different factors. The quality and quantity of available labour are critical. The type and size of trees, topography, forest soils and size of areas to be harvested all influence the choice of logging system.

Using eastern Canada as an example Boyd (1977) has listed most of the advantages and disadvantages of the forest scene there.

Disadvantages	Potential advantages
Forest is 80% in very small conifers.	Long, small, straight trees are suitable for multi-stem harvesting and bulk handling.
Much of the forest is on soft organic soils.	Potential development of high flotation vehicles suited to operations without road constructions.
Much of forest area is remote.	Very large operating areas suited to large efficient machines.
Much of the forest area is located far from established communities and labour force.	Minimal interference from alternate land use, environmental restrictions etc.
Slow tree growth.	Low land values and stumpage.
Low wood quality.	Suitable for bulk handling and relatively high fibre loss, use of wastes for fuel.
Long distance to mills.	Minimal interference from existing networks in developing efficient new transport systems.
High cost of road and railway lines.	Low land value, few right-of-way problems, large concentrations of wood to be moved.
Lack of communities, housing facilities.	Opportunities for planned communities, government aid.
No stable local labour supply, contractors or support service.	Opportunities for very large integrated operations and joint ventures under company control - use of minimum labour systems with very high labour rates of payment.
Mixed stands in many areas including many poor quality hardwoods.	Opportunity for full utilization, bulk handling, energy from wastes.

Based upon the above listing of perceived advantages and disadvantages of the forest regions in eastern Canada Boyd suggested some characteristics of harvesting systems that would find application there:

- multiple stem harvesting and processing, and bulk handling procedures are particularly suited to the small trees, low quality and mixed stands.

- the size of operating blocks is a major potential advantage and large highly productive machines are uniquely suited to the large Canadian situation.

- large operating units are feasible, including some realignment of areas to permit "swamp logging" units, "mountain logging" units, with equipment systems and personnel selected to suit the varying stand-terrain-soil situations.

- rights of way are available for permanent, innovative new transport systems handling large wood volumes, and joint ventures are possible. Government assistance and participation in opening up new areas are likely.

- conversion of low quality wood and residues for energy can be particularly attractive for Canadian forests. The utilization of such material can improve conditions for reforestation as well to convert the many silvicultural slums to vigorous growing stands.

- low quality and mixed stands suggest that systems with efficient centralized high-volume sorting and processing may be particularly effective.

- systems must have low labour input and may require unusually high wages for remote areas.

- systems minimizing the need for road building, operating off-road with a good level of efficiency are well suited to the situation.

- transport of labour force over long distance to and from large communities would be particularly desirable for Canadian remoteness. Some companies are transporting their workers by aircraft.

- harvesting systems which upgrade the future forest are desirable, eventually replacing the low wood quality of some areas by productive forest.

Figure 21 is a schematic representative of systems that utilize advantages of the Canadian forest industry situation (Boyd, 1977).

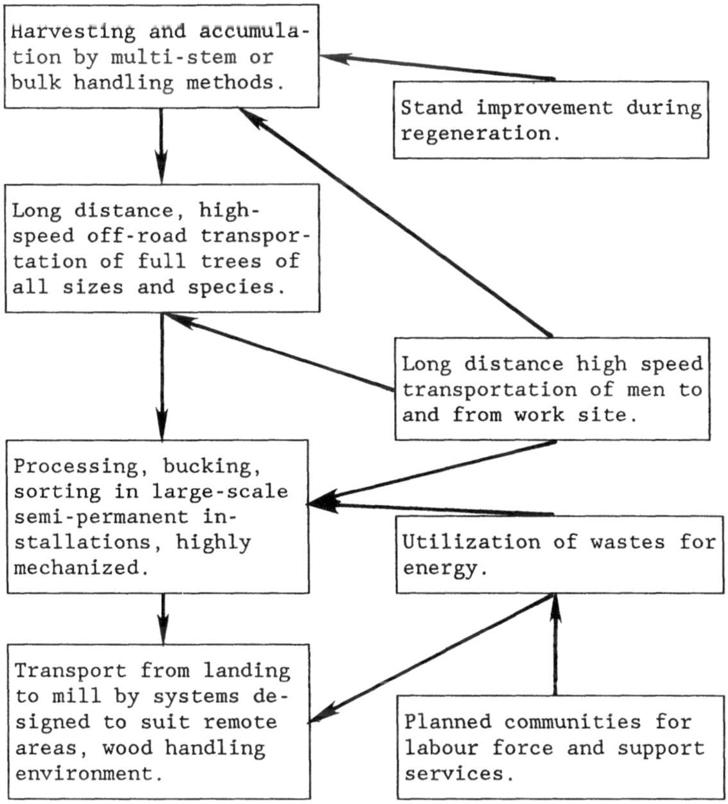

Figure 21. Systems that utilize the advantages offered by the Canadian forest resource (Boyd, 1977).

A similar approach to cataloguing the disadvantages and the perceived advantages of any forest region should prove valuable in establishing the direction in which to move when considering the harvesting of a new forest area. It does not provide the analysis necessary for a final decision but should establish the parameters for further detailed study. An example of this in a tropical region might produce a list as follows:

- mixture of species. The natural tropical forests are heterogeneous and it is common to find over a hundred different species on a few hectares. Many species are unknown with little if any knowledge of their properties. The assured economic supply of any given species is usually low (Sayn-Wittgenstein, 1972).

- only a low percentage of the total volume per hectare is harvested which increases logging costs. With few exceptions only 5-40 m^3/ha are extracted in the natural high tropical forest (Bendz and Järvholm, 1970).

- trees are on average much larger than in most temperate forests, in most instances 50-100 cm in diameter. The large size means that moving logs by human and animal power is difficult and often impossible. Many trees have high buttressed root systems which add to the difficulty in felling.

- there is a lack of good road-building material and soils in general have a low load-bearing capacity. In many regions fine sharp sand results in the rapid wear of steel tracks (Fellows, 1974).

- many tropical species have a high specific gravity and will not float. This precludes or greatly complicates water transport.

- unusual hardness or high silica content make some species difficult to fell and crosscut and chip.

- most tropical forest areas are remote from population centres and this results in poor communications and lack of services. Heat limits physical work capacity.

Nemtsov (1975) points out that the natural factors in a forest are uncontrollable and most of them have a probabilistic stochastic character, whereas the parametres of machines are controllable and can be modified within regional limits. For given machine parameters and available technology, productivity will depend then only on the natural factors and the limitations of human capabilities. The determination of the optimum parameters for harvesting machines for various working conditions requires a knowledge of the effect of the natural factors upon the machine parameters.

The same point was made by Bekker (1961) in a paper dealing with the evaluation and selection of optimum vehicle types under random terrain conditions. He pointed out the need for considering random terrain in terms of statistical probability while the specifications of a machine, tracked or wheeled, are fixed or deterministic.

6.1 EFFECT OF STAND AND TREE CHARACTERISTICS UPON OPERATIONAL EFFICIENCY

Methodologies for forecasting production of complete logging systems were developed in North America and Scandinavia during the later part of the 1960's. At that time two essential prerequisites for a systems approach were at hand. A complete data base on the rate of output of machines and machine components with quantified influences of site factors was available for the first time. Computers had become available which allowed quick and cheap computation of complex models and relationships. Examples are the studies at Skogsarbeten (Hedbring et al, 1966) and at the Canadian Forest Management Institute (McCraw and Silversides, 1970) and the work by Canadian/Swedish researchers on mechanized thinning (Newnham and Sjunnesson, 1969).

The Swedish methodology developed by Skogsarbeten was modified for application under North American conditions and found wide application there.

While the methodology was intended primarily to permit a comparison of various machines and harvesting systems under comparable forest stand conditions it also permits the user to analyze the effect of different stand and tree characteristics upon the harvesting process. Computer modelling and simulation studies have become very common in forestry and serve very useful functions. However studies such as this developed in Sweden enable one with pen and paper to make a quick comparison of forest stands and machine systems to establish the impact of forest stand conditions on productivity or to direct the operator towards the most effective machine size and type and harvesting system to employ.

Basic to the methodology is the break down of the various forest stand and machine characteristics into very small discrete units which can be altered to represent different stands and then to calculate the significance of any or each upon the productivity of the harvesting machines and system under consideration.

An example of this methodology is reviewed in Sections 6.1.1-6.1.4. Many similar production models have been published in recent years, many of them of greater complexity and sometimes with adaptions to the stochastic character of logging operation. However, this example will explain the general approach of analysis of this type.

The stand and machine characteristics considered are:

A = area in m^2/ha = 10 000
f(L) = number of cuts per tree - this will depend not only on tree height but also product to be produced

G = merchantable volume of tree (m^3)

H = number of tree lengths or full trees per set or in a stockpile on a landing (a set is the area in m^2 within which a machine may perform a particular function without moving)

k = a wander factor in the stand or on landings - the ratio between the straight line distance and the actual travel distance; e.g., straight line distance is 10 m but due to terrain and obstacles a machine must travel 16 m, then k = 16/10 = 1,6

Ld = number of trees per load

L_{lm} = merchantable length of longest tree in grapple load (m)

Lm = average merchantable length of tree (m)

L_{th} = total height of average merchantable tree (m)

M = merchantable volume (m^3/ha)

N = number of trees per grapple load

P = designated contents of pile or bunch (m^3)

s = straight line distance (m), can be distance between trees, or skidding distance

t = estimated fixed time/tree, for various functions (cmin = 0,01 min)

T = sum of the times of particular functions (T_1, T_2, T_3, etc.) of a machine (cmin)

v = travelling speed of any particular piece of equipment moving between trees, bunches of trees or logs in the stand, and along skid trails (m/cmin)

Wt = percentage of waiting time in processing cycle

Y = working area per set (m^2)

In the methodology as developed the productivity of each machine or harvesting system is expressed in m^3/h. An hour is 60 min or 6000 cmin. Calculated productivity must be modified by machine availability and efficiency of the whole operation to approximate actual operating conditions. These factors have already been discussed in some detail.

This methodology can be used to show the effect on productivity of stand volume, of tree size, of rough terrain and obstacles. It can also be used to compare different harvesting methods such as grapple versus choker skidders or the effect of tree size on skidding productivity.

6.1.1 EFFECT OF STAND VOLUME UPON FELLER BUNCHER PRODUCTIVITY. Feller buncher is mounted on a tracked chassis with rotating deck, articulated boom and circular-saw equipped felling head.

Variables:

k = 1,4; s = 5,22 m (@ 70 m^3/ha), 4,36 m (@ 100 m^3/ha), 3,69 m (@ 140 m^3/ha); t = fixed time per tree = 8 cmin; v = 0,3 m/cmin; M = 70 m^3/ha, 100 m^3/ha, 140 m^3/ha; Y = 25 m^2; G = 0,19 m^3; A = 10 000 m^2/ha

T_1 = Moving between sets

M = 70 m^3/ha

$$T_1 = \frac{A}{Y} x \frac{G}{M} x \left[\frac{k \; x \; s}{v} + t \right] = \frac{10\;000}{25} x \frac{0{,}19}{70} x$$

$$\left[\frac{(1{,}4 \; x \; 5{,}22)}{0{,}3} + 8 \right] = 31{,}0 \text{ cmin/tree}$$

M = 100 m^3/ha

$$T_1 = \frac{A}{Y} x \frac{G}{M} x \left[\frac{(k \; x \; s)}{v} + t \right] = \frac{10\;000}{25} x \frac{0{,}19}{100} x$$

$$\left[\frac{(1{,}4 \; x \; 4{,}36)}{0{,}3} + 8 \right] = 20{,}7 \text{ cmin/tree}$$

M = 140 m^3/ha

$$T_1 = \frac{A}{Y} x \frac{G}{M} x \left[\frac{(k \; x \; s)}{v} + t \right] = \frac{10\;000}{25} x \frac{0{,}19}{140} x$$

$$\left[\frac{(1{,}4 \; x \; 3{,}69)}{0{,}3} + 8 \right] = 13{,}5 \text{ cmin/tree}$$

T_2 = Felling trees within set	Estimated time
Swing and boom out to tree	4,0 cmin/tree
Position felling head at tree	8,0 cmin/tree
Sever tree from its stump	4,0 cmin/tree
Boom in, swing and drop tree or bunch	9,0 cmin/tree
	25,0 cmin/tree

T = $T_1 + T_2$

M = 70 m^3/ha = 31,0 + 25,0 = 56,0 cmin/tree = 20,4 m^3/h
M = 100 m^3/ha = 20,7 + 25,0 = 45,7 cmin/tree = 24,9 m^3/h
M = 140 m^3/ha = 13,5 + 25,0 = 38,5 cmin/tree = 29,5 m^3/h

The calculation of productivity, in m^3/h, is by the following formula:

$$\frac{\text{volume of tree } (m^3) \text{ x } 60 \text{ min}}{\text{total cycle time (cmin/tree) x } 100} = m^3/h$$

6.1.2 EFFECT OF TREE SIZE UPON FELLER BUNCHER PRODUCTIVITY. Feller buncher is mounted on a tracked chassis with rotating deck, articulated boom and circular-saw equipped felling head.

It is assumed that the stand volume per hectare remains the same but that tree volumes vary from 0,19 m^3 to 0,50 m^3. At 100 m^3/ha

Tree volumes m^3	0,19	0,30	0,50
Trees/ha at 100 m^3/ha	526	333	200
Area per tree, m^2	19,0	30,0	50,0
Tree spacing, m ($\sqrt{m^2}$)	4,4	5,5	7,1

Variables:

k = 1,2; s = 4,4 m, 5,5 m, 7,1 m; t = 8 cmin;
v = 0,3 m/cmin; M = 100 m^3/ha; Y = 25 m^2;
G = 0,19 m^3, 0,30 m^3, 0,50 m^3; A = 10 000 m^2/ha

T_1 = Moving between sets

G = 0,19 m^3

$$T_1 = \frac{A}{Y} \times \frac{G}{M} \times \left[\frac{(k \times s)}{v} + t\right] = \frac{10\ 000}{25} \times \frac{0,19}{100} \times$$

$$\left[\frac{(1,4 \times 4,4)}{0,3} + 8\right] = 21,7 \text{ cmin/tree}$$

G = 0,30 m^3

$$T_1 = \frac{A}{Y} \times \frac{G}{M} \times \left[\frac{(k \times s)}{v} + t\right] = \frac{10\ 000}{25} \times \frac{0,30}{100} \times$$

$$\left[\frac{(1,4 \times 5,5)}{0,3} + 8\right] = 40,4 \text{ cmin/tree}$$

G = 0,50 m^3

$$T_1 = \frac{A}{Y} \times \frac{G}{M} \times \left[\frac{(k \times s)}{v} + t\right] = \frac{10\ 000}{25} \times \frac{0,50}{100} \times$$

$$\left[\frac{(1,4 \times 7,1)}{0,3} + 8\right] = 82,3 \text{ cmin/tree}$$

T_2 = Felling trees within set

	Estimated time
Swing and boom out to tree	4,0 cmin/tree
Position felling head at tree	8,0 cmin/tree
Sever tree from its stump	4,0 cmin/tree
Boom in, swing and drop tree or bunch	9,0 cmin/tree
	25,0 cmin/tree

$T = T_1 + T_2$

G = 0,19 m³ T = 21,7 + 25,0 = 46,7 cmin/tree = 24,4 m³/h
G = 0,30 m³ T = 40,4 + 25,0 = 65,4 cmin/tree = 27,5 m³/h
G = 0,50 m³ T = 82,3 + 25,0 = 107,3 cmin/tree = 28,0 m³/h

The calculation of productivity, in m³/h, is by the following formula:

$$\frac{\text{volume of tree (m}^3\text{) x 60 min}}{\text{total cycle time (cmin/tree) x 100}} = \text{m}^3/\text{h}$$

6.1.3 EFFECT OF OBSTACLES AND ROUGH TERRAIN ON SKIDDER PRODUCTIVITY. It is assumed that because of a combination of slope and obstacles, a circuitous skidding route is required compared to a relatively straightforward trail. The wander factor k becomes 1.6 compared to 1.1 and skidder speed is reduced 50 percent. What effect does this have upon skidder productivity?

The skidder is wheeled, articulated frame-steered, equipped with winch, main cable and chokers - a one-man operation.

T_1 = moving between trees, machine stopped, man walking and choking trees.

Variables:

k = 1,1 and 1,6; s = 4,4 m; v = 0,24 m/cmin;
Ld = 3 m³ = 15,8 trees.

	Easy terrain	Difficult terrain
1.1 Dismounting from machine	1,3 cmin/tree	1,3 cmin/tree
1.2 Spooling out cable @ 0,24 m/cmin	6,2 cmin/tree	6,2 cmin/tree
1.3 Man moving between felled trees $\frac{k \times s}{v} = \frac{1,1 \times 4,4}{0,24}$	= 20,2 cmin/tree	—

$\frac{k \times s}{v} = \frac{1,6 \times 4,4}{0,24} =$	—	29,3 cmin/tree
1.4 Setting chokers	65,0 cmin/tree	65,0 cmin/tree
1.5 Mounting skidder	1,3 cmin/tree	1,3 cmin/tree
1.6 Winching tree lengths to skidder	5,7 cmin/tree	5,7 cmin/tree
	99,7 cmin/tree	108,8 cmin/tree

T_2 = Travel time to and from landing or roadside.

Variables:
k = 1,1 and 1,6; s =300 m; v = 0,8m/cmin and 0,4m/cmin;
Ld = 3 m^3 = 15,8 trees.

Easy terrain

$$T_2 = \frac{2(k \times s)}{v} \times \frac{1}{Ld} = \frac{2(1,1 \times 300)}{0,8} \times \frac{1}{15,8} = 52,2 \text{ cmin/tree}$$

Difficult terrain

$$T_2 = \frac{2(k \times s)}{v} \times \frac{1}{Ld} = \frac{2(1,6 \times 300)}{0,8} \times \frac{1}{15,8} = 152,0 \text{ cmin/tree}$$

T_3 = Offloading at landing

3.1 Release mainline, drop load and move ahead	10,0 cmin/tree
3.2 Dismount	1,3 cmin/tree
3.3 Unchoke tree lengths	25,0 cmin/tree
3.4 Mount skidder and winch in chokers	6,7 cmin/tree
	43,0 cmin/tree

	Easy terrain	Difficult terrain
T_1	99,7 cmin/tree	108,8 cmin/tree
T_2	52,2 cmin/tree	152,0 cmin/tree
T_3	43,0 cmin/tree	43,0 cmin/tree
	194,9 cmin/tree	303,8 cmin/tree

Productivity

Easy terrain $\frac{0,19 \times 60}{194,0 \times 100} = 5,85\ m^3/h$

Difficult terrain $\frac{0,19 \times 60}{303,8 \times 100} = 3,75\ m^3/h$

This calculated productivity must be modified by machine availability and efficiency of the operation to approximate actual operating conditions.

6.1.4 EFFECT OF TREE SIZE UPON WHEELED SKIDDER PRODUCTIVITY. It is assumed that the skidding distance remains constant, that stand volume is constant, the one variable being tree size which varies from 0,19 m³/tree to 0,5 m³/tree.

The skidder is wheeled, articulated frame steered, equipped with winch, main cable and chokers, one-man operation.

T_1 = moving between trees, machine stopped, man walking and choking trees.

Variables:
k = 1,4; s = 4,4m (0,19m³/tree) and 7,1m (0,50m³/tree); v = 0,24m/cmin; Ld = 3m³= 15,8 trees and 6,0 trees.

	Small trees	Large trees
1.1 Dismounting from machine	1,3 cmin/tree	3,4 cmin/tree
1.2 Spooling out cable @ 0,27 m/cmin	6,2 cmin/tree	15,5 cmin/tree
1.3 Man moving between felled trees		
$\frac{k \times s}{v} = \frac{1,4 \times 4,4}{0,24} =$	25,7 cm/tree	—
$\frac{k \times s}{v} = \frac{1,4 \times 7,1}{0,24} =$	—	41,4 cmin/tree
1.4 Setting chokers	65,0 cmin/tree	95,0 cmin/tree
1.5 Mounting skidder	1,3 cmin/tree	3,4 cmin/tree
1.6 Winching tree lengths to skidder	5,7 cmin/tree	14,0 cmin/tree
T_1	105,7 cmin/tree	172,7 cmin/tree

T_2 = Travel time to and from landing or roadside.

Variables:
k = 1,1; s = 300 m; v = 0,80m/cmin; Ld = 15,8 trees and 6 trees.

$\frac{2(k \times s)}{v} \times \frac{1}{Ld} = \frac{2(1,1 \times 300)}{0,8} \times \frac{1}{15,8}$	= 52,2 cmin/tree	——
$\frac{2(k \times s)}{v} \times \frac{1}{Ld} = \frac{2(1,1 \times 300)}{0,8} \times \frac{1}{6}$ =	——	137,5 cmin/tree

T_3 = Offloading at landing

3.1 Release mainline, drop load and move ahead	10,0 cmin/tree	26,3 cmin/tree
3.2 Dismount	1,3 cmin/tree	3,4 cmin/tree
3.3 Unchoke tree lengths	25,0 cmin/tree	35,0 cmin/tree
3.4 Mount skidder and winch in chokers	6,7 cmin/tree	17,6 cmin/tree
	43,0 cmin/tree	82,3 cmin/tree
T = $T_1 + T_2 + T_3$	200,4 cmin/tree	392,5 cmin/tree
Productivity		
$\frac{0,19 \times 60}{200,4 \times 100}$	5,7 m^3/h	——
$\frac{0,19 \times 60}{392,5 \times 100}$	——	7,6 m^3/h

This calculated productivity must be modified by machine availability and efficiency of the operation to approximate actual operating conditions.

6.2 RANGE OF TREE SIZE

Tree size has an important effect on operational efficiency and production cost. Based upon many studies analyzing productivity and cost, productivity in harvesting has been found to vary directly with tree size, the larger the tree the higher the man-day productivity. The impact of tree size on production is lessened somewhat with some forms of mechanized harvesting but still remains.

The volume of a tree stem is a function of its diameter breast height squared. A tree with diameter of 12 cm DBH has half the volume of a tree 18 cm DBH and one quarter the volume of a tree 25 cm DBH. The cost and low productivity resulting from handling small trees one at a time is obvious. Productivity would have to be increased four times when harvesting 12 cm DBH trees as compared with 25 cm DBH trees and this would be difficult to do, handling one tree at a time. Figure 22 shows the effect of tree size on various stages of tree harvesting (Conway, 1977).

The impact of tree size is particularly noticeable in the skidding operation. Handling one tree at a time with increased efficiency will only maintain the same shape cost curves. Conway states that what is required is multi-stem handling machines rather than multi-function machines and concludes that only through the introduction of multi-stem handling machines can the shape of the cost curve be changed and the effect of tree size on harvesting production and cost be reduced.

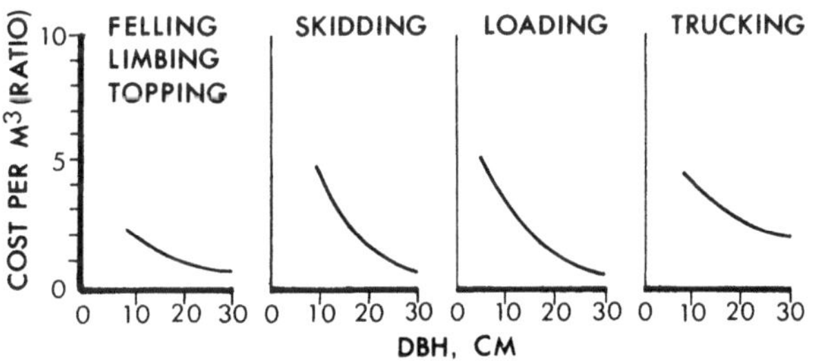

Figure 22. Effect of tree size on various stages of tree harvesting (Conway, 1977).

Escalating costs for increasingly smaller trees occurs in most logging systems. However the slope can vary and also the cost levels of the various sub-operations, depending on engineering design and system properties and also on the ratio between labour and machine costs. The curves in Figure 22 represent one logging system.

As manual felling and handling of small- to medium-sized trees are being replaced by harvesting machines, the impact of tree size on productivity can be reduced. This is the result of the development of felling machines capable of accumulating a number of trees prior to laying them down on the ground. Such developments have increased productivity up to 40 percent over manual felling (Sarles and Luppold, 1986). The cross-sectional area of the felling head has become a critical factor in felling productivity, almost as much as tree size. Figure 23 shows the increase of the feller buncher with an accumulator head compared to one which can fell and grapple single trees only. Such accumulator heads can result in a doubling of production with small diameter trees and with a substantial increase in productivity over single tree heads with larger trees up to a size that only permits two trees in the accumulator.

Another study is illustrated in Figure 24. It shows that for trees up to approximately 30 cm diameter breast height there is a very gradual increase in the time required to harvest with a Timbco feller buncher and with trees greater than 30 cm there is a rapid increase in the time required. With multiple stems harvested per cycle, the mean time per tree declines with the number of trees and increases very gradually with an increase in tree size.

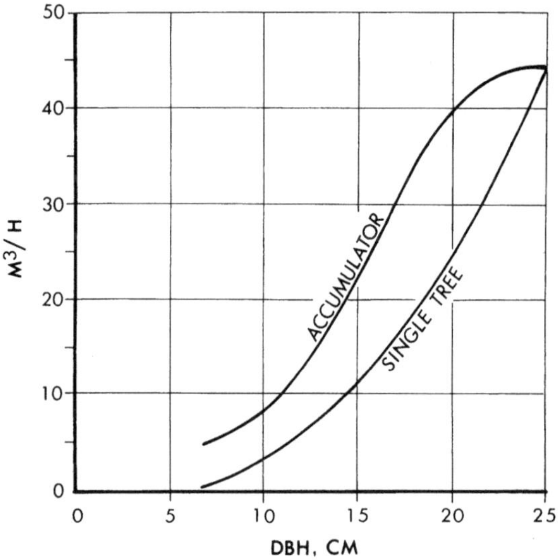

Figure 23. Comparison of shear productivity, accumulator versus single tree (John Deere 544 feller buncher with Rome accumulator shear).

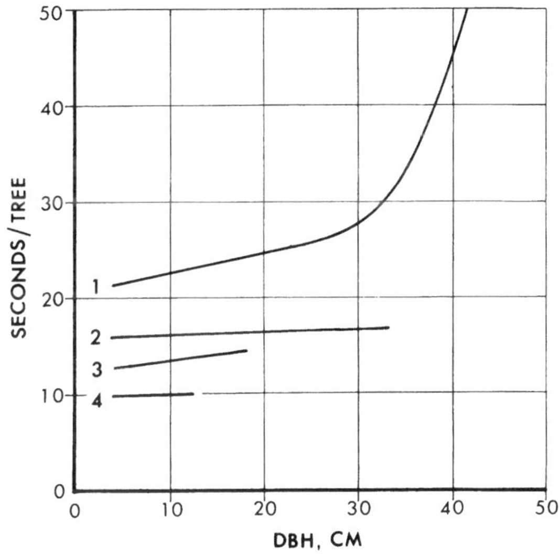

Figure 24. Mean fell and bunch time in seconds/tree versus DBH, cm for 1, 2, 3, 4 trees per cycle, Timbco system (Gonsior, 1986).

Recent developments in the design of accumulator heads provide a holding capacity of 0,56 m^2 or 6-8 trees to 20 cm in diameter. Such capacity is equal to or greater than the volume of a single tree 50 cm in diameter.

Besides increasing productivity in felling and collecting or bunching trees this operation impacts upon the skidding operation.

Figure 25 illustrates the possible effect of tree size upon harvesting machine productivity. This is an example only of harvesting machines. Others performing similar or different functions may have productivity curves of varying slope and shape. The Koehring Shortwood Harvester is a highly sophisticated single-tree harvester which fells, limbs, tops, cuts to length and forwards an 18 m^3 load of 2,44 metre length wood. The location and felling of a tree are controlled by the machine operator but once the tree is severed from its stump the processing operations of limbing, topping, cutting into lengths and storing the resulting bolts are all automated. In this respect the machine has advanced through mechanization to automation.

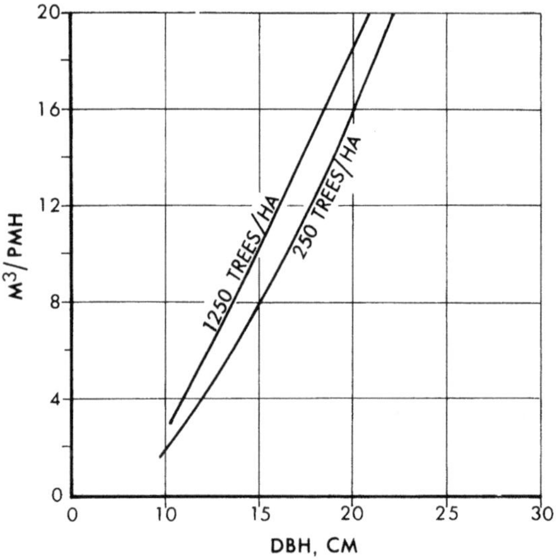

Figure 25. Machine productivity with various size trees, in m^3 per productive machine hour (m^3/PMH). Koehring Shortwood Harvester (Newnham, 1971).

The element of automation frees the operator to locate and fell another tree concurrently while the last one is being processed. When the carrier is full the harvester moves to roadside where it offloads itself into stockpiles or onto waiting trailers (Kurelek and Denovan, 1969). At an average forwarding distance of 200 m the time period from standing tree to bolts piled at roadside averages 1,2 minutes per tree harvested. Figure 25 also indicates the lesser effect of stand density on the productivity of this tree harvester. A five-fold increase in stand density, from 250 trees to 1250 trees/ha results in productivity increases up to 3 m^3/productive machine hour while a 5 cm increase in DBH results in productivity increases up to 8 m^3/productive machine hour.

A wide range in tree size such as occurs in natural all-aged boreal forests has a direct effect upon the productivity of felling machines as such machines are designed to handle maximum sizes of trees to be encountered but process trees of average size in the forest stand. The design engineer's maxim is "design for maximum but calculate performance on averages." For example, the maximum size of tree to be considered may be 50 cm DBH while the average merchantable stem may be only 20 cm DBH. Without the capacity to accumulate stems and if handling only one tree at a time, the machine will operate well below its capability most of the time.

Very large trees can only be felled manually (with chain-saws) and transported in the stand by cable systems or by heavy ground skidding equipment. Man-day productivity in this respect is high because of the volume of each unit (log or tree length) treated. A stand table showing the distribution of tree sizes, and if possible, a table indicating log diameters and lengths should be made to permit the selection of equipment suitable for the range of tree sizes to be harvested. If logs are plotted to show their distribution in percent by number and by volume it will be possible then to establish the portion of the logs of the total number or of the total volume either exceeding or lying below a given characteristic (Chardin et al, 1959) (Anon, 1967). From such curves (Figures 26 A and B) a selection of optimal sized equipment can be made.

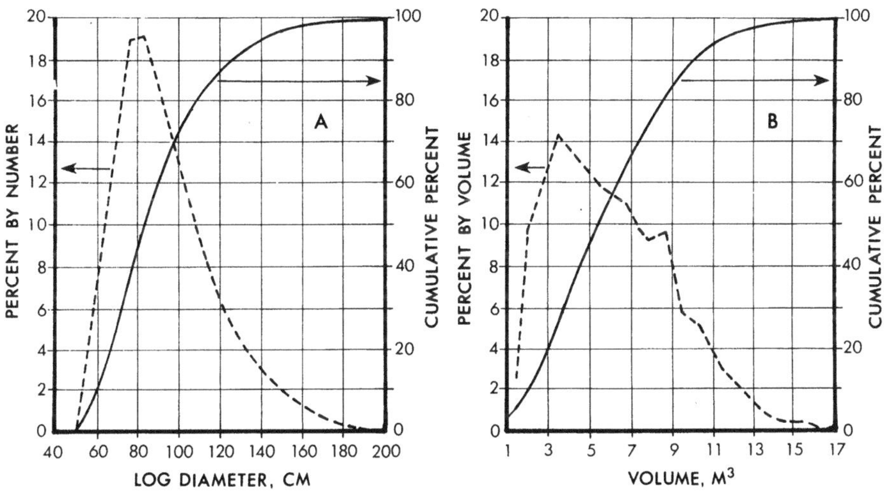

Figure 26. Distribution of log sizes, Ivory Coast, by log diameter (A) and by log volume (B). Average log length 6,5 m and average log volume 4,8 m³.

An example of the use of such data is given in Table 12.

Table 12. Numerical and volumetric percentage of trees with diameters exceeding the design one, in the major logging regions of the USSR (Nemtsov, 1975)

Regions	Dbh = 40 cm Number %	Volume	Dbh = 48 cm Number %	Volume	Dbh = 60 cm Number %	Volume
North	1,6	8,3	0,31	2,1	-	-
Centre	3,8	17,6	0,9	5,2	0,09	0,2
Urals	4,6	18,6	0,8	8,4	0,12	0,9
Siberia	12,3	47,6	8,9	29,0	2,4	1,1
Far East	16,7	51,8	9,7	36,6	3,8	18,5
All regions	7,4	30,2	3,5	17,8	0,5	3,5

For a machine designed to handle trees up to 48 cm DBH, in Siberia, 8,9 percent of the trees representing 29,0 percent of the volume would be too large. If the machine was designed to handle trees up to 60 cm DBH, then 2,4 percent of the trees representing only 1,1 percent of the forest volume would be too large.

If mechanized operations are designed into the logging method from the beginning it should be possible to maximize the output of any of the harvesting machines. It is often possible to generate a synergistic effect to increase the overall output, to the point where the whole is greater than the sum of the parts. Such an effect results, for example, in a machine which removes limbs being also capable of placing the tree lengths in a pile. It costs much less for a man with a chainsaw to fell and limb trees than it does with a tree harvesting machine. However the piled tree lengths can be handled by means of a grapple skidder whereas the manually felled and limbed trees must be handled by a choker skidder.

In analyzing the impact of tree size on the productivity of a tree processor where the tree is processed lengthwise from butt to top, the basic equation developed by J.A. McNally* is:

$$Pn = \frac{3600\ Vy}{\frac{(L-1)\ 60}{Yp} + tf + E_t}$$ in which

* McNally, J.A. Personal communication

Pn = normal hourly productivity, m^3/h
3600 = seconds in hour
V = tree volume, m^3
y = effective factor (including personal time but excluding availability) = 0,9
L = total length of tree, m
l = total length of tree above the butt which must be fed into the processor before processing takes place (2 m)
Yp = feed speed of processor, m/min
tf = time required to feed tree into processor, independent of Yp (4 s)
t = sum of all delay times
- operating delay 1,0 s/tree
- boom delay 1,5
- processor delay -
- relocation delay 3,0

5,5 s/tree

This equation was developed for the Logging Research Associates Arbomatik processor but can be applied to any processor such as the one- or two-grip harvesting machines with suitable modifications for the operations to be performed, i.e., if the operation is intermittent rather than continuous.

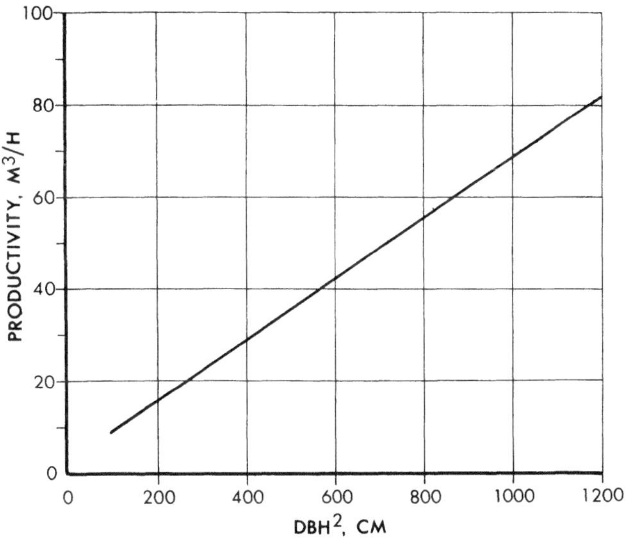

Figure 27. Relationship between productivity and tree diameter squared. Feed speed in example is 45 m/min.

A rule of thumb that is often used for processors of this type is that the productivity of the machine, for any given feed speed, will vary directly as the square of the tree's diameter at breast height. This rule of thumb applies to coniferous trees in the boreal forest. The rule is illustrated in Figure 27.

In the case of boom and grapple loading equipment, a wide variation in tree or log sizes and weights can be dealt with to a degree by reducing or extending the reach of the boom. Load capacity of boom in this example is in approximately inverse proportion to lifting radius, e.g., 2, 4, 6 m radius allows 4900, 3000, 1600 kg loads (Figure 28) (Anon, 1982).

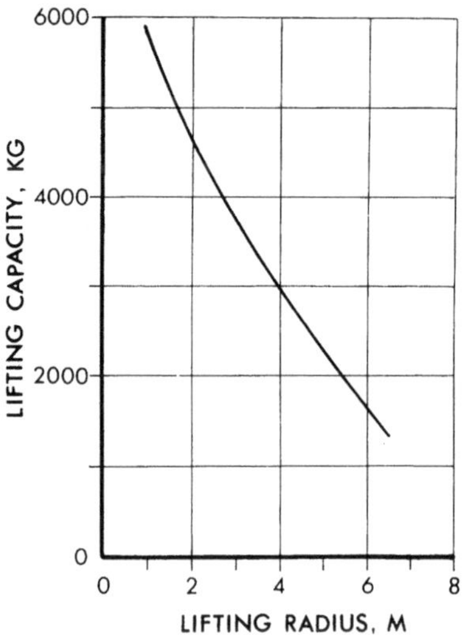

Figure 28. Load capacity of knuckle boom loader as function of lifting radius.

6.3 UNIFORMITY OF TREE SIZE

Uniformity in tree size such as is found in man-made even-aged forests results in higher productivity and more economical logging than in uneven-aged and uneven-sized forests. Uniformity or consistency is important in any production process, whether in timber harvesting or in manufacturing. Uniformity in size of trees permits the design and manufacture of machines which can be sized to the average size tree, with very little variation around the average.

The important parameter for machine design is probably the volume present in each diameter class, rather than the number of stems. It is obvious that there will be fewer stems of maximum size but they will have a substantial volume.

Forest stands of uniform size lend themselves to the production of uniformly sized products, pulpwood, sawtimber, poles, etc., and tend to simplify both the harvesting and the manufacturing processes.

Small diameter and very large diameter logs result in increased handling time, the former because of the extra motions involved, the latter because of the disproportionate amount of human energy required.

Given a stand of timber of a known total volume, and with a known mean diameter, the range of tree sizes may vary as is shown in Figure 29. The area under the curves is the same in each instance as it represents the stand volume of the total number of trees in the stand. It will be seen that as the base becomes narrower the height of the curve increases, now often described as kurtosis. The narrow base represents a close dispersion around the mean which would be representative of an even-aged stand. The height of the curve represents the numbers in each diameter class. As the height of the curve decreases and the base broadens the tree diameters, in number or volume, spread out from the mean. This is representative of an all-aged or uneven-aged stand where the range of diameter classes is greatly increased.

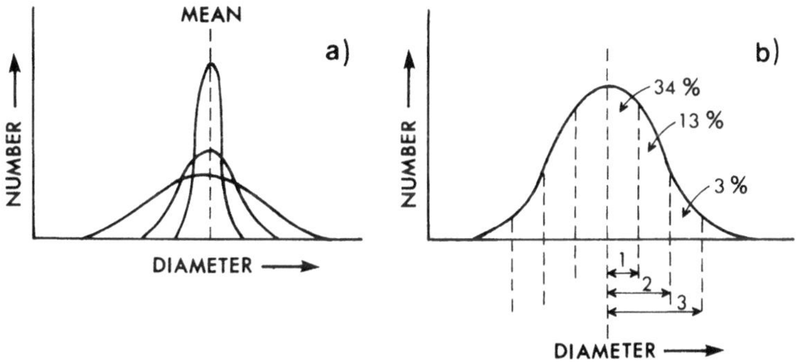

Figure 29. a) Various ranges of tree size around common mean diameters for stands of same volume. b) Standard deviation showing that 68% of stand lies within 1 standard deviation from mean and 94% of stand lies within 2 standard deviations.

The most important measure of dispersion is the standard deviation. This is useful in establishing for any forest area the sizing of equipment as it will indicate the range of sizes distributed around the mean.

The standard deviation is found by the following equation:

$$s = \frac{(x - \bar{x})^2}{n}$$ in which

s = standard deviation

x = number of trees in each measured diameter class

$\bar{x}$ = mean or average diameter for all trees in stand (or sample)

n = total number of trees in stand (or sample)

The distribution of tree sizes can also be shown in cumulative curves as in the Figure 30. Such distribution curves are useful in the sizing of harvesting machines which can handle trees up to a maximum size, as the curve will tell what quantity will fall outside the range of the machine's work capability.

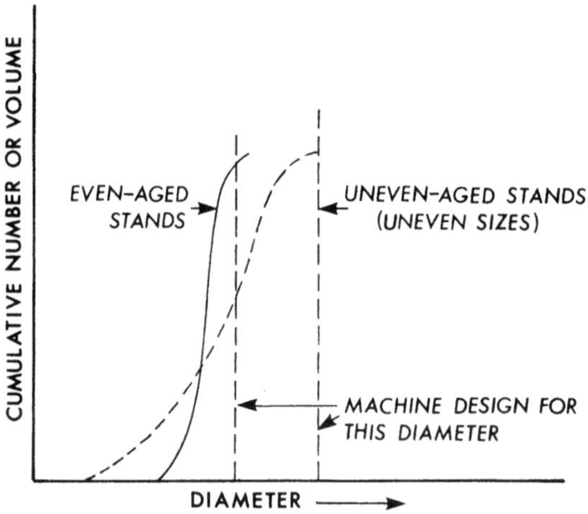

Figure 30. Use of cumulative curves to assist in sizing harvesting machines.

The curves shown in Figures 29 and 30 are called normal distribution (or Gauss distribution) and are symmetric around the mean. However, most forest stands have an asymmetric distribution of diameters, with the top of the curve dislocated to the left. Several mathematical formulas are published which attempt to fit such distributions but they will not be reviewed here. Only some more simple ways to quantify the properties of the distribution curves will be mentioned, using Table 13 as an example of a typical spruce stand in Canada.

Table 13. Example of an asymmetric distribution of trees for a typical spruce stand in Canada

(1) Diameter breast height, cm	(2) Number of trees
10	22
12	28
14	33
16	42
18	50
20	47
22	42
24	33
26	30
28	21
30	18
32	16
34	15
36	12
38	10
40	7
42	5
44	3
46	1
48	0
50	1
	420 (N)

First, the <u>median diameter</u> can be established, cutting the number in two equal halves. In Table 13, there are 420 trees and the median will be just below 20 cm as 210 trees are smaller and 210 trees are larger. The median can be supplemented by the lower and upper quartile, cutting off 1/4 of the numbers. The lower quartile is located in the 16 cm diameter class as 105 trees are smaller and the upper quartile in the upper range of the diameter class 28 cm as 105 trees are larger.

Thus:		
	Median diameter	approx. 19,8 cm
	Lower quartile	< 16,5 cm
	Upper quartile	> 28,4 cm

To be as precise as possible one should also know how the diameter classes were established. In Table 13 the 20 cm diameter class can consist of all trees between 20,00 and 21,99 cm or of all trees between 19,00 and 20,99 cm depending how measurements and recording originally were made. The difference between the two measurements is 1 cm which could be of significance in the sizing of machines.

A second assessment is the arithmetic mean diameter, computed from the formula

$$\bar{d} = \frac{\Sigma d_i \cdot n_i}{N} \quad \text{in which}$$

$\bar{d}$ = arithmetic mean diameter

d_i = diameter class i

n_i = number of trees in diameter class i

N = total number of trees.

Arithmetic mean diameter is often used in the computation of work studies on the cutting of trees as the total time consumption for felling the trees has proven to be well correlated to the mean diameter.

The diameter of the tree of mean basal area is often used in silvicultural research, computed with the formula

$$\bar{d} = \sqrt{\frac{\Sigma d_i^2 \cdot n_i}{N}}$$

For the distribution in Table 13 we get:

Median diameter	19,8 cm
Arithmetic mean diameter	23,1 cm
Diameter of the tree of mean basal area	24,1 cm

6.4 SPATIAL DISTRIBUTION OF TREES

Tree spacing varies widely within and between forest regions, between species whether tolerant or intolerant, or a mixture of the two. Tree spacing influences harvesting to varying degrees, whether clear cutting, thinning or selective cutting. Clear cutting as the term is commonly used does not always mean the complete clearing of a forest area. It is usually applied to certain merchantable trees of acceptable quality and species. Unacceptable trees, perhaps 10-25 percent of the original stand are usually left standing. In selective cutting only acceptable trees are removed, usually of good quality and desired species, and the residual stand left behind may form 60 percent of the original stand. In thinning, up to 70 percent of the original forest may be left.

Harvesting machines tend to move slowly in the forest, felling trees and dropping them in bunches for future skidding, or felling and processing trees into bolts or log lengths and piling them down or transporting them within the stand to roadside. Such machines range back and forth within the stand harvesting trees and traversing much of the forest floor. Speed of movement is not so much a factor as is the capability to move over difficult and often hostile terrain. For this reason many such machines are usually mounted on tracks. Simulation studies by Newnham (1968) on tree harvesters, such as feller-buncher machines with a boom-mounted felling head, have shown that slewing speed and maximum reach are the most important variables in such machine concepts affecting tree-harvesting time. The spatial pattern of trees, whether uniformly spaced, random, or clumped, has little effect upon machine productivity (Figure 31). It should be noted that trees within a forest stand cannot be truly random since all distances between trees do not have an equal chance of occurring. These studies also indicated that the number of trees per hectare has little effect upon machine productivity.

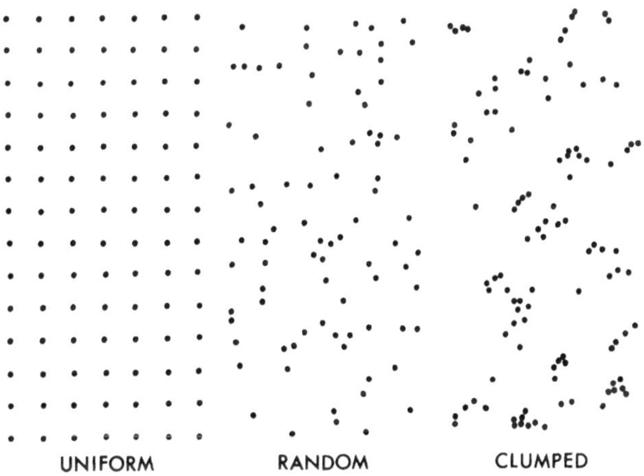

Figure 31. Spatial patterns of trees in different stands a) uniform, b) random, c) clumped (Newnham, 1971).

The movement of harvesting machines in the forest stand is different from that of forwarders or skidders whose purpose is to travel light and loaded, back and forth, transporting trees or tree sections to roadside. With few exceptions these machines travel over prepared or partially prepared trails. Speed is a relatively important factor with these machines as this is a major factor governing their productivity. For this reason, with a few exceptions, such machines are normally wheeled.

In the case of cable (choker) skidders, with average distances up to 150 m the speed of travel of the machine does not have a large impact on its total cycle time and productivity. However, with greater distances and for grapple skidders, travel time assumes an even greater portion of the total cycle time and machine speed is increasingly important.

Table 14. Grapple skidder versus choker skidder (Sarles and Luppold, 1986)

	cmin/tree Choker skidder (1)	Grapple skidder (2)
1. Moving between trees, machine standing still, man walking and choking trees (15 trees)	101,3	-
2. Loading bunch of tree lengths (20 trees)	-	19,6
3. Travel time to and from road or landing	49,4	36,3
4. Offloading at landing	43,0	5,0
5. Total time, cmin	193,7	60,9

In column (1) travel time is 26 percent of total cycle time while in column (2) travel time is 60 percent of the total cycle time although the machine speeds are equal.

The effect of skidding logs or tree lengths over a prepared or bulldozed trail compared to skidding over an unprepared forest floor is shown in Table 15 (McCraw and Hallett, 1970).

Table 15. Skidding tree lengths over bulldozed trails and unprepared trails (McCraw and Hallett, 1970)

Wheeled skidder class	Bulldozed trails Number of trips	Loaded km/h	Light km/h	Unprepared trails Number of trips	Loaded km/h	Light km/h	Mean load m^3
Medium <56 kW	386	4,7	6,2	555	4,0	5,3	2,22
Large >56 kW	659	5,3	6,9	73	3,2	4,0	3,65

The impact of prepared skidding trails is most evident with large skidders moving large loads, particularly the large clambunk skidders.

In the extraction of timber the construction of trails for the transport of such wood is a one-time, short-term event. After harvest such trails are filled in with regeneration or other growth. For this reason, only minimum monies should

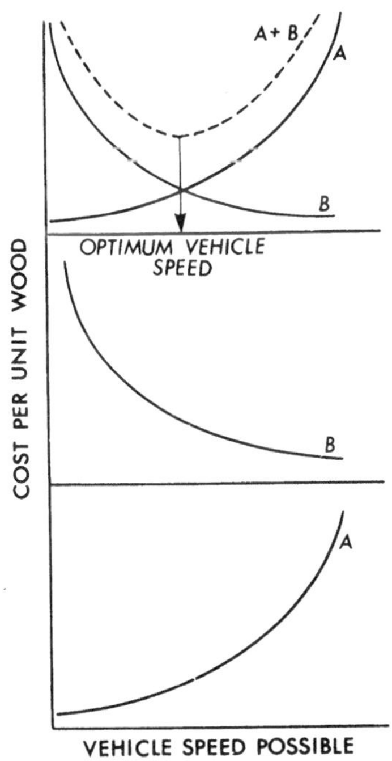

Figure 32. Minimum skidding or forwarding cost as related to speed; graph A trail construction cost, graph B direct cost of transport, top graph total cost with optimum speed.

be spent on such trails as will permit vehicle speeds which result in minimum transport costs (Figure 32).

The width, alignment and roughness of the trails will depend in large part on tree spacing in the stand. This in turn may depend upon the cutting practice, whether clear cut, partial cut or thinning. Sometimes a single pass with a tractor and bulldozer may suffice, in other instances the grubbing of stumps and roots and the filling of depressions or the cutting down of humps will be required.

Trail width, to permit ready passage without serious damage to bordering trees, will be related directly to machine width and length and trail curvature.

It has been said that any object moving in a given medium tends to assume the form which would offer the least resistance to motion. This tendency is borne out in nature as well as in the world of technology. In the case under consideration the medium or the environment is the forest, at its most difficult the forest in its virgin condition and at its easiest perhaps when the forest has been clear cut, with a wide range of conditions between these two extremes.

Few forest machines have been designed with an emphasis on their working environment. As has been stated most economic successes have been ecological failures. Only recently has the impact upon environment been seriously considered by machine designers.

Certainly ground clearance of machines was an early consideration because of its effect upon machine mobility but length and width and the general configuration of the forest machines depended more upon the manufacturers' production capabilities than on an effort to mate the machine to its environment.

On the premise that forest machines will be working in a forest stand and not be operating in the open without obstructions, such machines should be constructed with no projections to catch upon and damage residual trees. They should be designed to result in minimum damage to the soil and to residual trees while manouvering in the forest. The weight of the machine should be as minimal as possible within the constraints of design and manufacture and the weight should be distributed to result in a minimal ground pressure by the wheels or tracks.

Obstacle avoidance is aided by agility, the ability of the machine operator system to change direction rapidly and radically, and by keeping the vehicle narrow, its minimum turning radius small, and its swept width during a turn to a minimum (Nuttall, 1967).

In the original design of articulated wheeled skidders the centre of articulation was approximately equidistant from the front and rear axles. This configuration resulted in a minimum turning radius as the rear wheels followed precisely the track of the front wheels and being short-coupled, the sweep of the rear end of the machine was kept to a minimum. In some models and with articulated frame-steered forwarders and harvesters the rear frame is lengthened and with the longer wheelbase the point of articulation is closer to the front axle than it is to the rear axle. With the drive on both axles the inside rear wheel follows a shorter path than the inside front wheel when the vehicle is turning. If the wheels are being driven at the same speed, the rear wheel tends to overtake the front wheel and this causes the outside wheel to be lifted. The inside rear wheel must slip or drag otherwise it transfers weight from the outside rear wheel (Martin, 1965). Figure 33a shows a comparison between two machines, one with the point of articulation at a centre point between front and rear axles, the second with an extended wheelbase and the point of articulation ahead of centre (Bruun, 1986).

$$R_F - R_R = \frac{(H - 2Q)\ (1 - \cos\alpha)}{\sin\alpha} \qquad \text{in which}$$

R_F = turning radius for front wheel - m

R_R = turning radius for rear wheel - m

H = wheelbase - m
Q = distance from front axle to point of articulation - m
α = angle of articulation

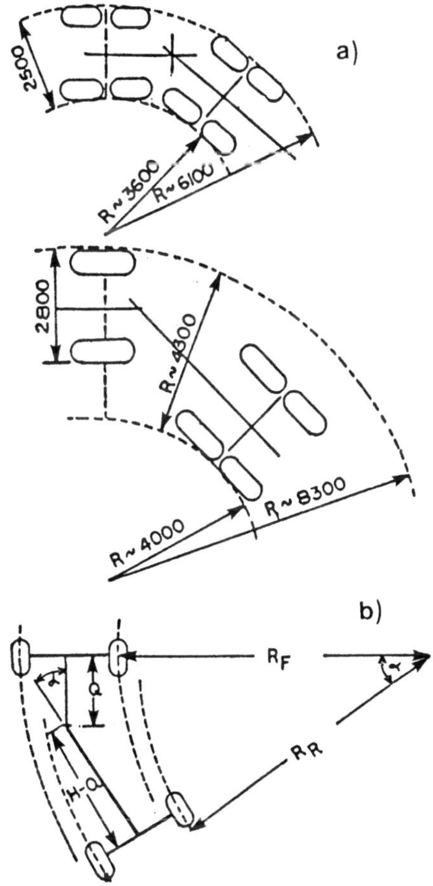

Figure 33. Turning patterns a) Machine with point of articulation at centre and with extended wheelbase and point of articulation ahead of centre; b) Influence of location of point of articulation on wheel slip, at various turning angles.

If in the example the machine wheelbase is 2,92 m and the point of articulation is 1,19 m behind the front axle, the difference in turning radius between the front and rear inside wheel will be 0,18 m, for an articulation angle of 35°. (Figure 33b)

In the situation as described it is evident that all 4 wheels on the machine are turning in unison but the rear wheels are pushing while the front wheels are braking. It becomes plain that much of the force generated in the ground contact by the wheels does not contribute to machine movement but tends to be destructive of the ground surface (Bredberg, 1984).

A wheel or bogie passing over an obstacle on one side of a machine will cause the unit to tilt and this increases the width of pass of the vehicle, compared to its operation on a flat unobstructed surface. The effect of such an obstacle is greater for vehicles with large diameter single wheels than those equipped with a tandem bogie (Figure 34).

In raising the side of a forwarder and tilting it the effect may not be great on the front of the machine if it is equipped with a walking beam or pivoting front axle but the side tilt can increase the operating width of a machine by up to one quarter (Håkansson, 1977). Figure 35 illustrates how obstacles in either or both directions impact upon the operating width of a machine. On a 10° slope up or down the tilt can increase the operating width of a forwarder by up to 36 percent (Lampen et al, 1970).

Mobility of a machine may be defined as the ease with which it moves about over a given terrain. As a measure of mobili-

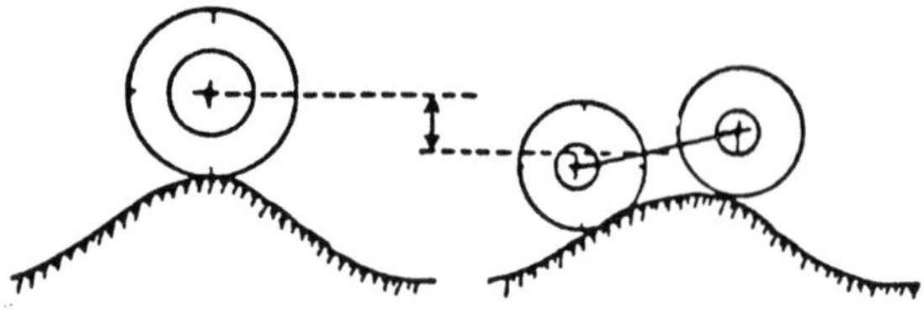

Figure 34. Obstacle raised side of vehicle with large single wheel higher than if equipped with tandem bogie (Bredberg and Wästerlund, 1983).

ty of a machine, speed within operation tolerance is important. One measure of vehicle performance has utilized the nautical term "speed made good." Speed made good is defined as the straight-line distance from the beginning to the end of the vehicle trip divided by the time required for the vehicle to make the trip. The increase in distance travelled as a result of manouvering during a trip is expressed in terms of a ratio of the actual trip length divided by the straight-line distance. This is sometimes referred to as the "wander factor." If, for example, the straight-line distance is 200 m but a vehicle is required to travel 250 m, then the wander factor would be expressed by 1,25.

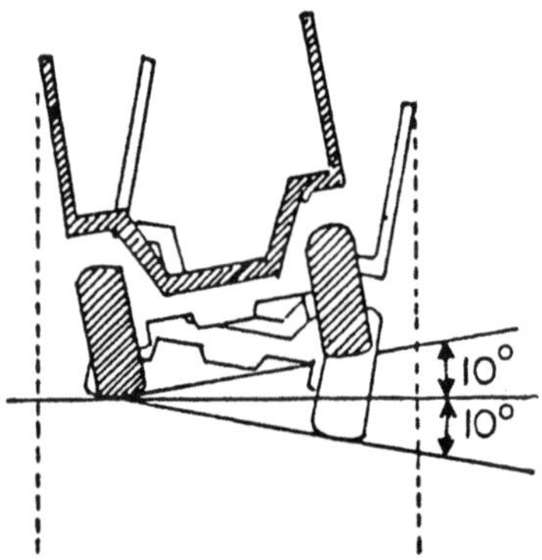

Figure 35. Effect of slope and or obstacles upon the operating width of a forwarder.

Off-road vehicle speeds are governed by vehicle vibrations caused by ground roughness, by physical weakness of the soil or by the necessity of bypassing impenetrable obstacles such as trees. The spacing and distribution of tree stems or stumps influences vehicle movement in a forest stand, at least to the extent that as the density of the stand increases, movement is hampered until at some stand density movement is denied. The effect of stem spacing upon the passage of a vehicle is considered to be a function of the characteristics of the vehicle, most commonly its width and turning radius.

For a stand of uniform and ordered distribution such as a plantation, spacing of the trees is implicitly understood as the distance from one tree to the next nearest. What does spacing mean in a natural forest where the distance between trees is variable and the distribution random to a degree? "True" randomness is rare in a natural forest.

The average spacing of objects in random distributions has been expressed by the following equation:

$$S = 0{,}6 \cdot \frac{A}{n} \qquad \text{in which}$$

S = spacing of obstacles, m
A = area of forest under consideration, m^2
n = number of trees

For example, in a forest area 1 km^2 there are 1 000 000 m^2 (100 ha). Assume a stand density of 800 trees/ha, then

$$S = 0{,}6 \; \frac{1\,000\,000}{800 \times 100} = 0{,}6 \times 12{,}5 = 7{,}5 \text{ m}$$

It is not the average distance or the minimum distance between trees or obstacles which govern the passage of a vehicle. It is the greatest distance between the obstacles and the arrangement of these into unobstructed paths (Addor, 1963). If a plantation is considered, spacing may be very little greater than the width of a forwarder or other vehicle but by following the "unobstructed path" passage is achieved with little difficulty. If, however, alternate rows are shifted, passage is denied even though the minimum spacing has not been changed (Figure 36).

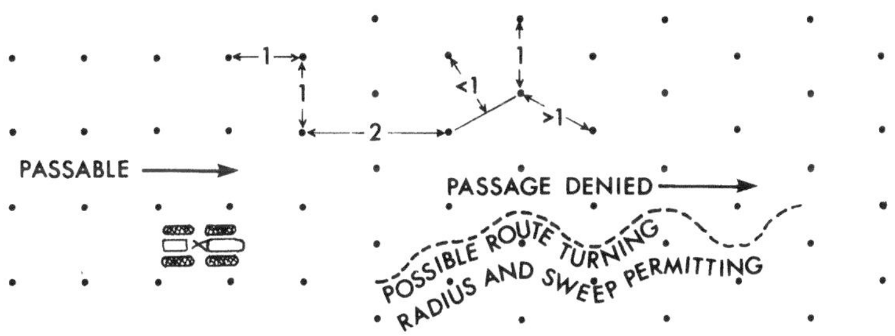

Figure 36. Effect of rearranging obstacles, minimum spacing unchanged (Addor, 1963).

During a turn the actual width of the vehicle is not a measure of the minimum passable distance between obstacles. The sweep of the vehicle, the actual area covered by the vehicle during a turn increases the effective width to something greater than the actual width.

The minimum obstacle spacing required for many vehicles is 1,4 times the vehicle width (Figure 37). This does not hold true for all and vehicle length and turning radius will affect the minimum tree spacing negotiable.

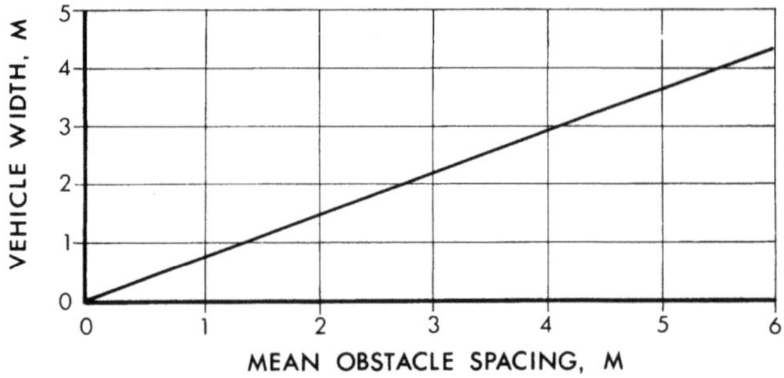

Figure 37. A vehicle can usually tolerate a mean spacing of obstacles down to 1,4 times its width.

The spacing of trees or obstacles has a direct effect upon the speed of travel within a forest stand, variations in forest soils perhaps to a lesser degree. Operator reaction to actual and anticipated obstacles has a major impact. Operators will drive vehicles at relatively high speeds on roads or well-prepared trails because they do not anticipate obstacles ahead. In a forest stand, with reduced visibility and much uncertainty as to distribution of obstacles and variation in ground conditions an operator will generally reduce the speed of his vehicle below its capability for travel, in relation to his perception of the obstacles to be encountered.

Figure 38 gives an indication of possible speeds obtainable in relation to the width of the vehicle and the spacing of obstacles. (Blackmon and Stoll, 1968).

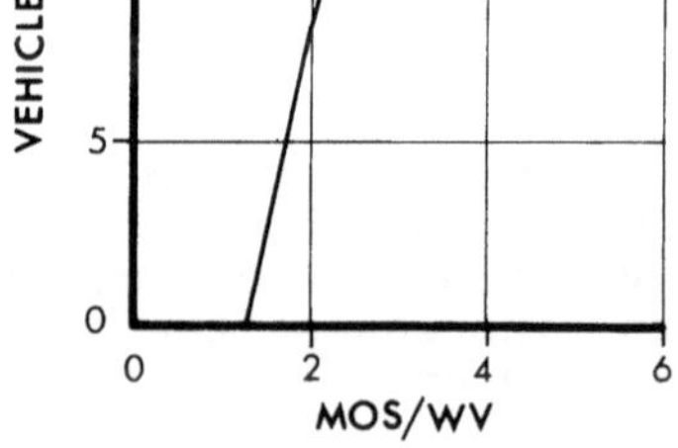

Figure 38. Vehicle speed in km/h as related to the ratio: mean obstacle spacing (MOS) to width of vehicle (WV).

A given area divided by the number of trees in the area results in an average area occupied by one tree. Considering this area as a circle with the tree in the centre, the radius of the circle considered is the distance that is free before entering the mean area of another stem. Since the mean area per tree is equal for all trees in the given area then the mean distance or spacing between the stems is equal to the diameter of the circle encompassing the mean area per tree. This diameter may be termed the mean obstacle spacing and it approximates the difference between tree centres.

Mean obstacle spacing may be calculated by diameter or area:

Diameter

$$\bar{A} = \frac{A}{20}$$

$$\frac{\pi d_a^2}{4} = \frac{\pi}{4} \cdot \frac{D_c^2}{20}$$

$$d_a^2 = \frac{D_c^2}{20}$$

$$d_a = 0{,}224\ D_c$$

and by definition $d_a = s_e$

Area

$$\bar{A} = \frac{A}{20}$$

$$\frac{\pi d_a^2}{4} = \frac{A}{20}$$

$$d_a^2 = \frac{4}{\pi} \cdot \frac{A}{20}$$

$$d_a = 0{,}252\ \sqrt{A}$$

$d_a = s_e$

A = given area studied = $\frac{\pi D_c^2}{4}$

$\bar{A}$ = mean area per stem = $\frac{\pi d_a^2}{4}$

D_c = diameter of structural cell

d_a = diameter of mean area per tree

s_e = mean tree spacing

The number 20 in the equation $\bar{A} = A/20$ reflects the concept of the structural cell (Mills, 1962). Using circular plots of increasing diameter around a central point the average area per tree is calculated. At some point the average area will approach an approximately constant figure if plotted over the radius of the sample. When this plateau point is determined the area so outlined is designated as the structural cell. In the study in question it was found that the structural cell exists with 20 individual trees. By counting 20 trees and establishing the area of the circle including them it is possible to develop mean tree spacing.

6.5 TREE CHARACTERISTICS

The physical characteristics of trees have an important impact upon the relative efficiency of harvesting. Coniferous trees, for example, characteristically have a single tapered bole with small diameter branches while many deciduous trees have a relatively short bole with a widely spreading crown, branched trunk and large limbs, many as thick as the tree bole itself.

Tree form in deciduous trees is commonly interpreted in terms of terminal bud dominance. In trees with excurrent branching the main stem or terminal bud out-grows the lateral branches resulting in cone-shaped crowns with clearly defined central boles. Trees with weak terminal bud dominance are said to exhibit decurrent or deliquescent branching with lateral buds growing almost as fast as the terminal stem and the central bole becomes lost in the central crown (Figure 39).

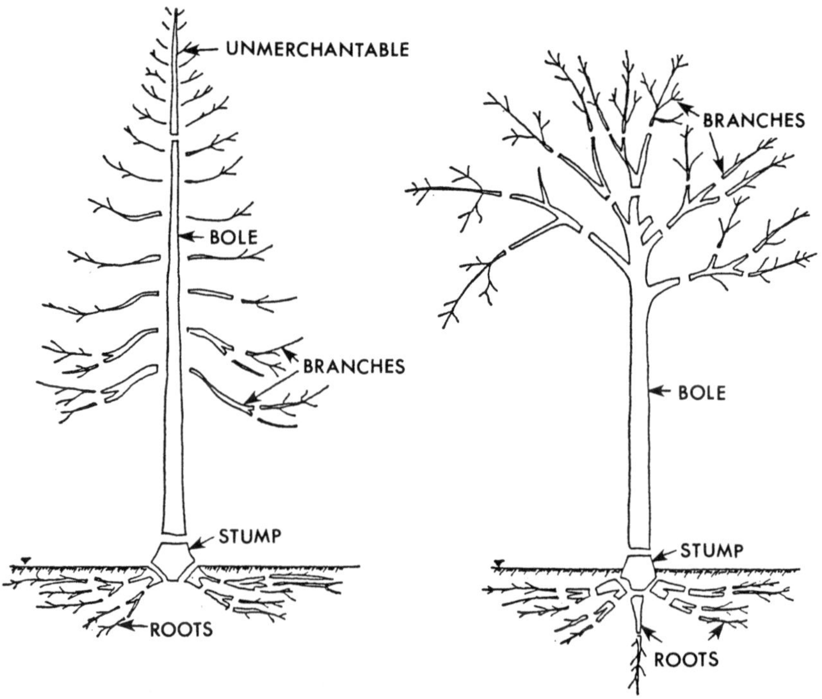

Figure 39. Physical characteristics of coniferous and deciduous trees.

The effect of these differences is immediately obvious when felling trees. The conifer with narrow crown and tapered bole can usually be felled with a minimum disturbance to surrounding trees. The centre of gravity of the tree is relatively low and the inertia of the falling tree is not great. With large spreading crowns deciduous trees can cause considerable damage to trees in the surrounding stand and often the trees will be hung up when felled. Further processing of the tree becomes difficult and dangerous in such an event. If the branches on the underside of the crown do not break they will support the stem and the upper branches well above the ground. As limbs are removed from the crown the weight distribution and stresses in the top are changed resulting in a tendency for the top to roll off the branches that are propping it up. The direction and the amount of roll is difficult to anticipate and the worker must be constantly alert (Gabriel et al, 1974).

With coniferous trees, small limbs and top leave relatively little residue when processed into logs and bolts in the stump area. The larger limbs and crowns of deciduous trees on the other hand usually result in a large accumulation of residue.

Table 16. Relative above-ground portions of coniferous and deciduous trees (Thomas, 1978)

	Coniferous	Deciduous
Stemwood (including bark)	80%	73%
Branches and foliage	20%	27%
	100%	100%

Note: Above data for eastern Canadian conifers and tolerant hardwoods.

Because of the nature of industrial development in the northern hemisphere the mechanization of tree harvesting has concentrated almost entirely upon conifers, because of their principal physical characteristics - a straight single main bole with a gradual taper and small branches.

A few deciduous trees can be handled mechanically, such as intolerant species like Betula and Populus which have long clear stems and relatively small crowns. Even these species, because of their inherent branching characteristics, are sufficiently irregular to cause problems.

6.5.1 TREE BRANCHING HABITS. Branches have a considerable impact upon operational efficiency in logging. As pointed out by Hakkila (1971) crown length and limbs to be removed affect product productivity; wage scales in forest work are determined in part by the branchiness of trees; productivity of branch chippers depends upon the length and thickness of branches; in whole-tree logging and transport the additional weight of green branches is a major factor. Large branches in the crowns of deciduous trees tend to jam in the spout of chippers and have to be cut by saw or axe to enter the chipper. With some machines, limbs less than 10 cm in diameter can be crushed by feed rollers. Chipping delays are often a function of branch angle along with branch diameter (Herschelman and Domenech, 1980).

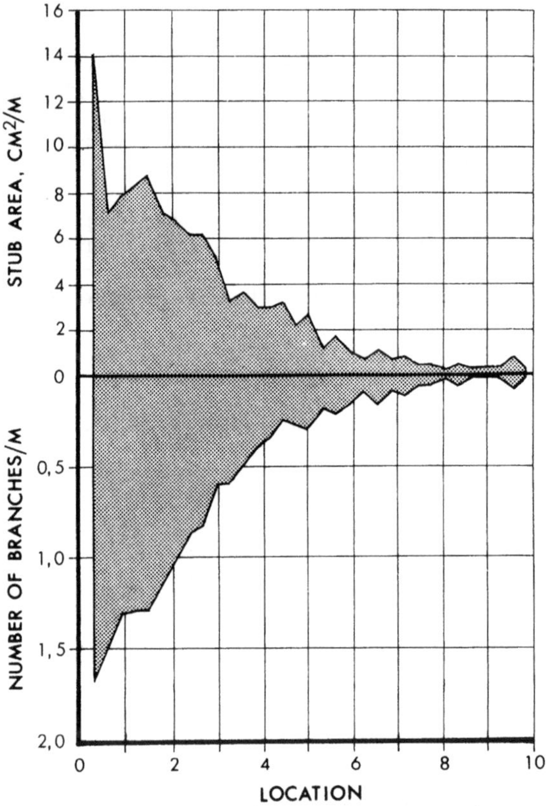

Figure 40. Average number of branches and average stub area in cm²/m section of the bole located at various distances downward from the top cut at a diameter of 7,5 cm, jack pine (P. banksiana).

It has been established that in pine and spruce in eastern Canada the live branches of a tree are not uniformly distributed longitudinally along the bole. This is particularly noticeable in the pines where the branches are concentrated in whorls at the nodes of the stem, the internodes being branchless or with only a few small branches present. The distribution of branches along the bole of a tree is shown in Figure 40. It describes jack pine and shows the number of branches and the branch stub area in square centimetres for an average tree, measuring downwards from the height at which the diameter of bole is 7,5 cm (merchantable top diameter). It is seen that there is a steady decline in the number and branch stub area, with an increase in distance below merchantable height (Drolet et al, 1971).

As a tree grows in a closed stand its lower branches become shaded and as a result many trees drop such limbs and become self-pruning. This is clearly shown in Figure 41 as the lower portion of such trees has a section of bole free of branches, a section of bole which supports dead branches which have not yet dropped and in the upper portion of the crown which is exposed to sunlight, live growing branches.

Tree species vary widely in tolerance to shade and in turn will be affected by stand density of the crown canopy and the shade cast by it.

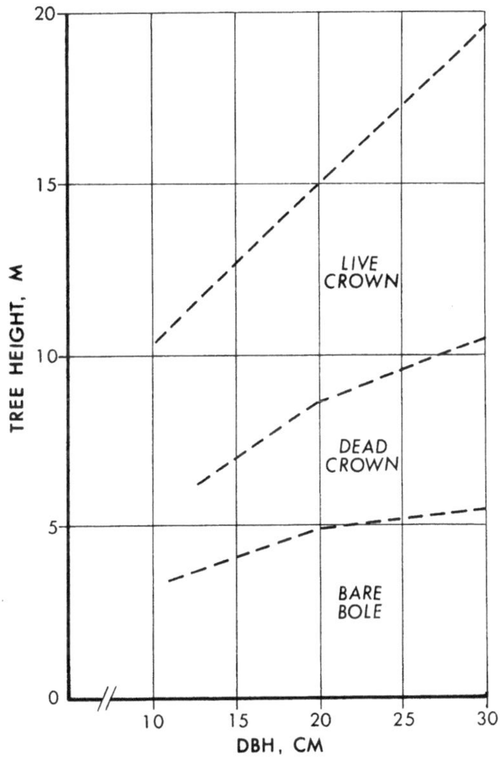

Figure 41. Distribution of tree crown for jack pine.

Because of the impact that limbiness has on productivity in harvesting, trees and/or forest stands may be classified to indicate the ratio of live crown to bare stem (Figure 42). The impact of limbs on productivity in harvesting can be illustrated by the development in Sweden of the Logma delimber in the 1960's. This machine was unique among delimbing machines as it picked up and limbed a tree from its top, towards its butt. The logic here was that only that section of the tree supporting branches was processed. Other machines operating from the butt of a tree towards its top had to process the whole merchantable bole length including that section of the tree with no branches. Another argument in favour of this technique was that the Logma only had to limb a tree from its merchantable top diameter down to the base of the live crown or lower if there were many persistent dead limbs, and didn't have to accommodate the largest part of the tree, its butt. For this reason the capacity of the machine could be scaled downward so that, for example, the delimbing mechanism capable of handling a diameter of 32 cm could in fact process a tree with a 50 cm butt diameter.

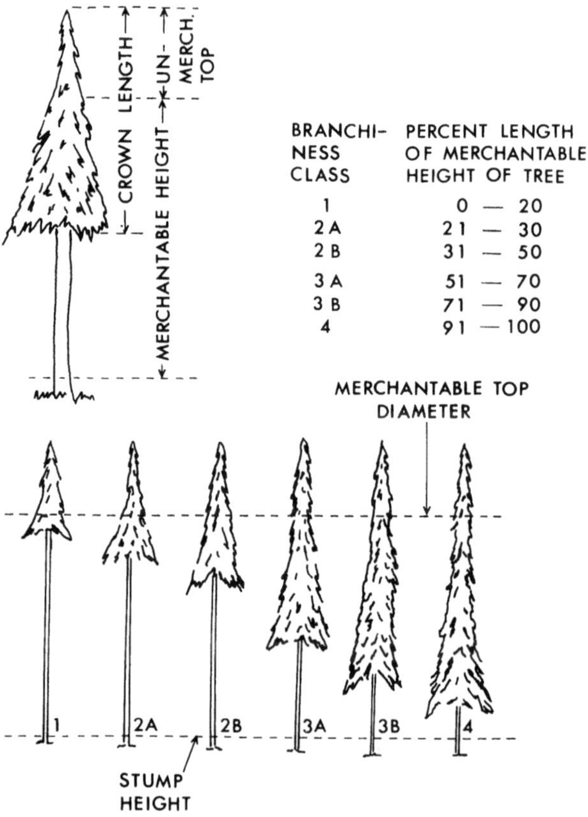

Figure 42. Example of the classification of trees by crown length.

The diameter of branches of coniferous species, measured at their juncture with the tree bole, varies with species and the variation within species varies in turn with the size of the tree. Figure 43 shows the variation in branch diameter for Scotch pine (P. sylvestris) and Norway spruce (Picea abies) in southern Finland (Hakkila, 1971). The limbs on pine are considerable larger than those of spruce for trees of the same diameter breast height. Ager (1972) found that with both pine and spruce there was a rapid decrease in branch diameter in the first 5 cm outward from the surface of the tree bole. The decrease is as great as 20 percent of the branch diameter.

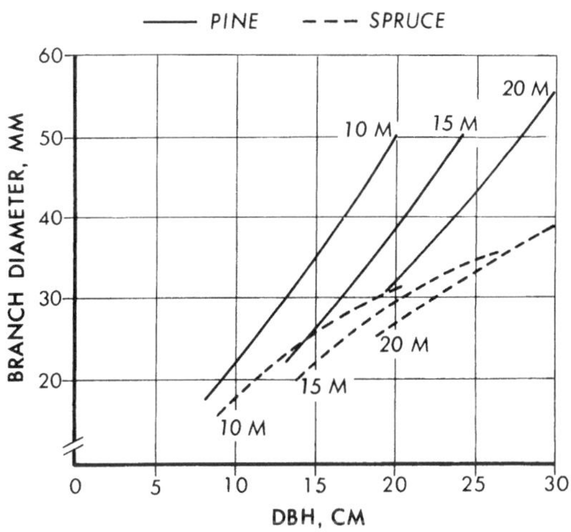

Figure 43. Variation in branch size in relation to tree diameter for Scotch pine and Norway spruce in southern Finland (Hakkila, 1971).

The density and strength of branchwood are greater than that of stemwood (Brown et al, 1952). The average density of branchwood in Norway spruce in south Finland was found to be 151 percent of the density of the stemwood in the same tree while for Scotch pine it amounted to 105 percent (Hakkila, 1971).

When full trees are being piled on the ground or loaded onto transport it is the branch stub, up to 10 cm in length, that has the most influence on the density of piles. Branches break off from the stem under such conditions and when trees are piled together the spaces formed by the branch stubs are enough to accommodate flexible limbs of green and unfrozen trees (Chinn, 1982). With loblolly pine (P. taeda) branches make up 10 to 15 percent of the total tree weight depending upon tree size. The proportion of small (<1,4 cm), medium (1,5-4,9 cm) and large branches (>5 cm) varies dramatically with tree size (Taras et al, 1974 and Clark et al, 1976). The proportion of large branches increases rapidly with increasing tree size while small to medium branches show a steady decline (Figure 44). Yellow poplar (Liriodendron tulipifera) in the southern United States shows a proportionate decline in smaller branch size with increasing tree size (Figure 44). The pattern of branch size followed that of the pines, branch size increasing with an increase in tree diameter (Clark and Schroeder, 1977).

The length of branches is also variable between species and there appears to be a direct relationship between tree size and branch length. Figure 45 illustrates this for Scotch pine, Norway spruce and black spruce.

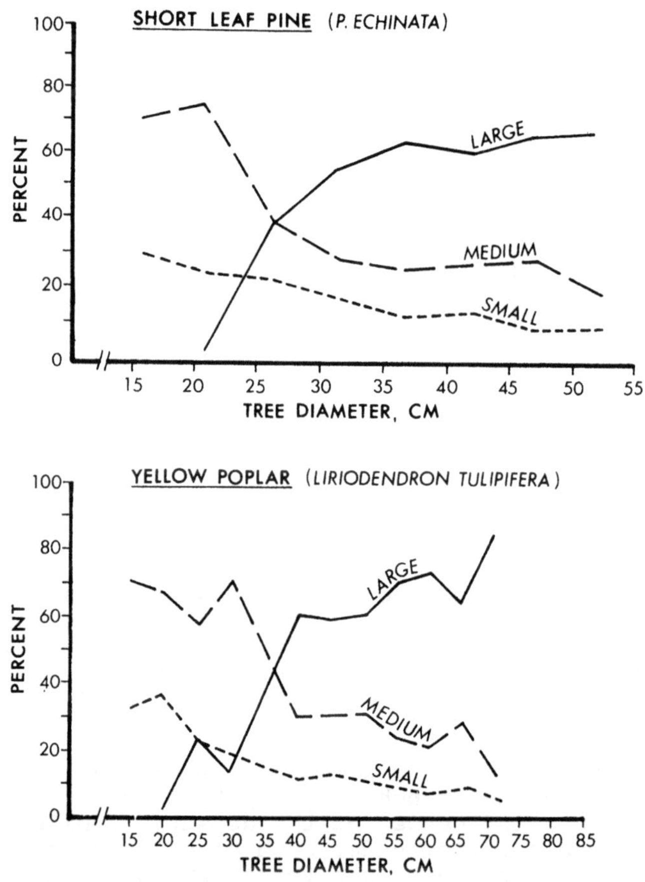

Figure 44. Distribution in percent of small (≤1,4 cm), medium (1,5-4,9 cm) and large (≥5 cm) branches in relation to tree size for short leaf pine (P. echinata) and yellow poplar.

A study made in eastern Canada found that branches are not evenly distributed around the circumference of a tree but that in eastern Canada the southwest quadrant of the trees supported the largest number of branches and the northeast quadrant the least (Drolet et al, 1971). Even when the quadrants for each tree were re-ordered so that quadrant one, regardless of aspect, always contained the greatest number of branches, nearly 50 percent of the total branches for jack pine and 40 percent for black spruce and balsam fir, the quadrant diametrically opposite contained half to one third the number of branches (Figure 46). This uneven distribution of branches will influence the density of piles of full trees as it should result in closer packing of trees than if the distribution of branches circumferentially was regular.

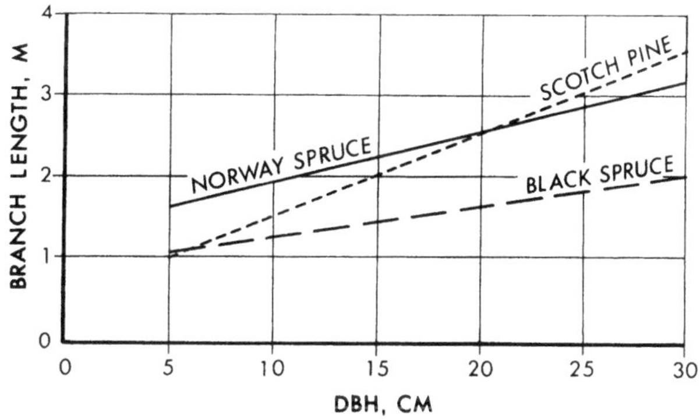

Figure 45. Relationship between branch length and tree size.

Besides dealing with branches on individual trees, crown volumes of trees in the aggregate are also important. This is particularly true in whole-tree harvesting where trees may be debranched at a landing or concentration point. Accurate figures on the density of such piles of branches to relate weight to volume or vice versa are scarce and would require checking. The Pulp and Paper Research Institute of Canada, in a study, established an average weight per cubic metre of such limbs of 40,0 kg (Keen, 1963). An earlier study in the U.S.S.R. (Elpatievsky et al, 1934) developed the figures in Table 17.

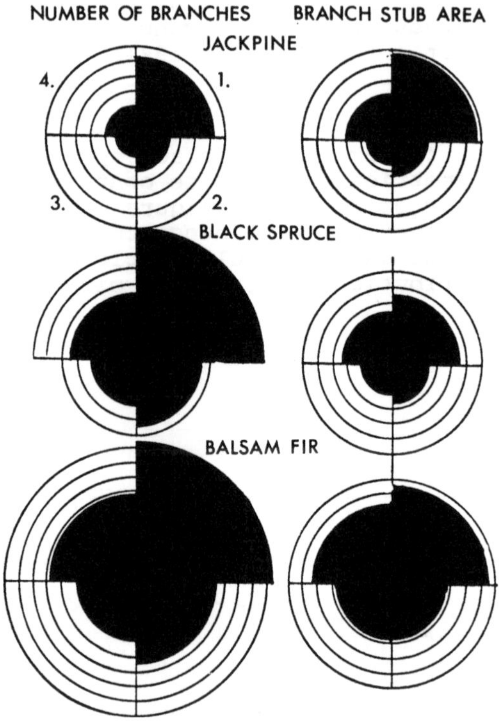

Figure 46. Distribution of branches and branch stub area around tree stem per 0,305 m section of crown. Distance between successive circles represents one branch or 6,5 cm² (one square inch) of branch stub area. (Drolet et al, 1971).

Table 17. Coefficients of density of slash piles

Condition	Species			
	Spruce		Pine	
	Coeff.	Density	Coeff.	Density
Loose	0,3	(19,2 kg/m³)	0,4	(25,6 kg/m³)
Average	0,5	(32,0 kg/m³)	0,6	(38,4 kg/m³)
Compact	0,8	(51,2 kg/m³)	0,9	(57,7 kg/m³)

These figures for average density are close to those developed by the Pulp and Paper Institute of Canada.

7. FOREST SOILS

Timber harvesting is primarily a transportation operation. A knowledge of forest soils and their characteristics with regard to passing traffic, both wheeled and tracked, is critical to efficient operation.

Foresters are normally concerned with site quality, soil nutrients, etc. rather than with the engineering characteristics of soils which affect machine trafficability and mobility. Soil trafficability is the capacity of the ground to sustain the traffic of vehicles. It will vary with the ability of the soil in question to withstand the effects of the horizontal shearing forces and the vertical loads imposed upon it by wheels and tracks. Mobility of a vehicle is defined as the absolute ability of a vehicle to traverse a specific route or a specific type of terrain (Thomson, 1968).

The soil is a complex system made up of solid, liquid, and gaseous material. The solid portion may be mineral or organic, the liquid portion is the soil water which fills part or all of the open spaces between the soil particles. The gaseous portion occupies that part of the pore space between the soil particles that is not filled with water.

In general classification forest soils have been grouped into the following categories:

7.1 FINE GRAINED SOILS

These include silts and clays varying in particle size from 0,005 mm to 0,05 mm. Silt is a finely ground rock flour with barely visible grains. Particles feel gritty in the mouth. Silt has little or no plasticity and no cohesion. Care is required to distinguish silt from very fine sand and from clay. If mixed with water silt particles will settle out in from 30 minutes to one hour. Clay is a fine inorganic material consisting largely of dust along with a fine colloidal fraction which is for the most part hydrous aluminum silicate derived as a product of chemical weathering. It is both cohesive and plastic so that it can be rolled into a long thread when moist. It forms hard lumps when dry, often impossible to crush with the hands. It does not feel gritty when bitten with the teeth. It will remain suspended in water 3 hours to indefinitely.

The strength of these soils is determined by their moisture content as water controls the inter-particle adhesion. At about 12 percent moisture content on a dry weight basis such soils reach their maximum strength. As the soil moisture increases, water fills the interstices between the soil particles and the cohesion which existed between them declines. Above 16 percent moisture content the soil particles become immersed, surface tension disappears and the soil mass becomes a viscous liquid with little shear strength.

7.2 COARSE GRAINED SOILS

Particles varying from fine sand (>0,5 mm) to coarse sand (<2 mm) to gravel (>2 mm) are considered as coarse-grained soils. They are normally not as sensitive to water as are the fine-grained soils although some moisture in sand will act as a binder and improve its trafficability. Sand is granular and gritty. Grains are loose and may be readily seen and felt. Sand lacks plasticity and cohesion although it may show considerable stability when wet, due to surface tension phenomena. The stability of moist sand is demonstrated by the use of tidal flats for automotive traffic before drying has progressed too far. Gravel generally consists of rounded pebbles but sometimes of angular rock grains. It is gritty and granular and crunchy underfoot. It has no plasticity or cohesion. It has been arbitrarily defined as having a grain diameter up to but no more than 15 cm. It is rarely lacking mixture with other soils. "Bank run or pit run" gravel is a term usually applied to a mixture of sand and gravel, often with an appreciable clay or silt content (Figure 47).

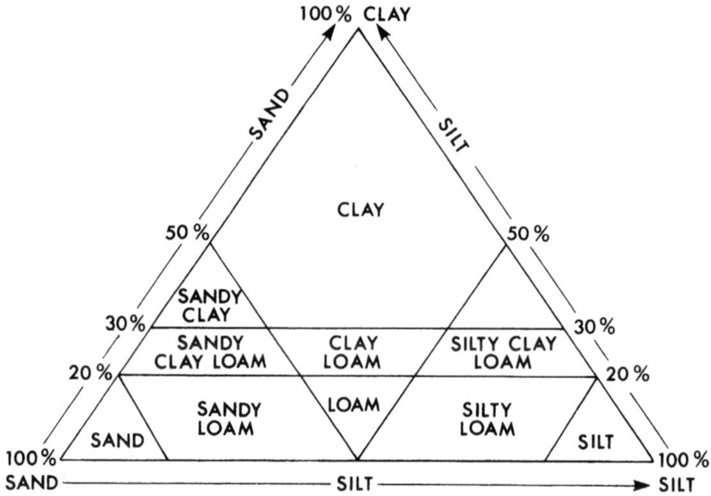

Figure 47. Chart indicating combinations of basic soils, sand, clay, and silt.

7.3 ORGANIC SOILS

These may be divided into four broad groupings: forest floors, muskeg, marsh and tillable humus. All groups are characterized by low shear-strength and high water retention. Forest floors may be considered as a special form of muskeg, the difference being the depth of the organic layer. Forest floors are an accumulation of decaying organic material interlaced with living root systems. In the natural state the moisture content is governed by the vegetation growing there, except perhaps in depressions. Where the vegetation has been removed the water content may increase and the trafficability of the soil decrease.

Muskeg surfaces may be covered with lichens, mosses, shrubs or trees. The roots of these plants extend through the underlying mass of humus and provide what strength there is to this type of organic soil - the more woody roots the better the trafficability of the muskeg. Care must be taken to preserve the root mat structure by limiting the number of vehicle passes.

Peat consists of partially decayed plant material. It is fibrous in nature and plant remains are usually visible in it. Because this form of soil is to be found around the world where forests occur and because of the problems it presents to harvesting it will be described in some detail. Plants during their growth take carbon, hydrogen and oxygen from the air to form cellulose. Plant debris at the end of the growing season falls to the ground. Where it is exposed to the action of air, oxidation takes place rapidly and a great part of the organic material is resolved into gases, the carbon for the most part being converted to CO_2. If the debris falls into water or upon soil which is saturated with water and atmospheric oxygen is excluded, decay proceeds very slowly. A large portion of the fixed carbon is retained, hydrogen and oxygen forming water (H_2O), carbon and oxygen forming carbon dioxide (CO_2) and carbon and hydrogen forming methane (CH_4).

As decomposition proceeds the vegetable matter is converted into humic and allied acids, and carbonic acid, while the soluble compounds slowly pass into solution and the resulting mass contains a large and increased proportion of carbons. The character of the peat depends upon the conditions under which it is formed and the nature of the vegetation which has contributed to its formation. A comparatively small number of plant species supply the bulk of the content of muskegs - mosses, grasses, rushes, sedges, erica and vaccinium, and sometimes the trunks, roots and leaves of trees.

Muskeg has a great capacity for taking up and holding water. Water is considered a component of muskeg as much as is the peaty material (Radforth, 1952). Most bogs contain 90 percent moisture. The high water-holding capacity of peat results from its high colloidal properties. Peat is an irreversible colloid. Drying will destroy its colloidal properties and the resulting dry substance consists of aggregates of associated particles which are no longer capable of absorbing large quantities of water.

Marsh is organic terrain too wet to support large woody growth. Usually marshes are found where bacterial action proceeds almost the year round and organic debris is digested rather than accumulated. Humus will pack quite firmly under the action of moderate traffic particularly if incorporated with clay or sandy loams. However if water is allowed to gather in traffic-caused ruts, the soil is readily weakened.

7.4 ASPECTS OF SOIL STRENGTH

In logging transportation foresters are primarily concerned with the capacity of a soil to withstand traffic of tracked or wheeled vehicles, or trafficability. This will vary with the ability of the soil in question to withstand the effects of the horizontal shearing forces and the vertical loads imposed upon it. Soil strength, over and above the minimum required by the vehicle for self-propulsion, is required to provide the excess traction the vehicle must develop to overcome other terrain obstacles.

The two basic soil properties that define trafficability are interrelated as the bearing capacity of a soil is considered to be a function of shear strength, and to a lesser degree, cohesion. A shear stress will tend to displace particles of an earth mass with respect to each other or to displace a part of the earth mass with respect to the remainder. There is a slip of particles if they are displaced with respect to each other. This displacement of particles or parts of the earth mass is opposed by its shearing resistance. This shearing resistance is passive and develops only if there is a shearing stress and the value of shearing resistance equals the value of the stress at any time. As the shearing stress increases so does the shearing resistance up to the point where the resistance of the earth mass is exhausted and failure occurs. This point is the shear strength of the soil.

Figure 48 is taken from "Relation between soil and a vehicle" (Bekker, 1950). Let us consider two soil particles. The lower particle is immobilized and the upper particle is loaded with a weight W. The force F required to move the upper soil particle in the direction shown by the arrow is:

$F = W \cdot \tan \emptyset$ in which
$\tan \emptyset$ = the coefficient of friction between the two surfaces

Experience has shown that the angle Ø for shear resistance for clay may vary from 0 - 20 and for sand any value up to 35 depending upon the moisture content of the specimen.

The shearing stress is equal to a normal pressure multiplied by a coefficient (Bekker, 1950).

In b the soil particles are not bound together by mechanical friction but by a cohesive binder. If no load W is applied to the upper soil particle, the force F required to break the loaded area would be:

$F = c \cdot A$ in which
A = the contact of the area of the surface
c = the cohesive strength of the binder, per unit area

By combining the forces in a and b, the general formula (Coulomb's Formula) for shearing an average soil which combines both mechanical friction and cohesion is obtained:

$$F = W \cdot \tan \emptyset + c \cdot A$$

In applying these formulae it is assumed that what applies to two particles of soil will also apply to the whole soil mass. Therefore the shear strength of a soil depends upon two distinct properties which complement each other. Depending upon which property is predominant, soil strength is more or less sensitive to loadings and contact areas applied.

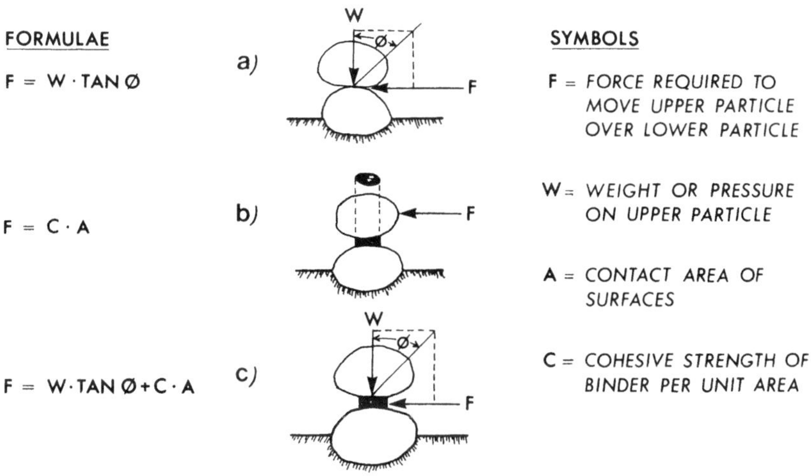

Figure 48. Soil strength. a. Mechanical friction between soil particles, b. Cohesion between soil particles, c. Friction and cohesion acting together.

7.4.1 COHESION. Cohesion in soils may be of two kinds, a true cohesion caused by the natural attraction of particles due to molecular forces or an apparent cohesion due to the action of moisture films. True cohesion in soils is infinitesimal and can be disregarded.

An example of apparent cohesion may be observed when dry fine sand is moistened. The sand becomes cohesive as the surface tension of the water acts at the surface of the water films on each grain and holds the sand grains together. However if all the air is replaced by flooding with water there will be no air to water surfaces at which surface tension can act. Sand flooded with water has no cohesion but tends to flow like a liquid. Thus there is a certain proportion of sand to water which gives maximum cohesion and optimum moulding properties.

Grain or particle size also affects cohesion. If two equal volumes of sand are considered, each wetted to give maximum cohesion but each of a different grain or particle size, the sand of smaller grain size will contain a greater number of surface tension bonds and will therefore have greater cohesion. For example there is practically no cohesion in a pile of small pebbles, no matter what amount of water is used to moisten them.

Table 18. Coefficients of cohesive resistance (c) and angle of internal friction for various soils (Anon 1929, Hogentogler, 1937)

Material	kg/m²	ø (degrees)
Silts - wet	0	10
Clay - liquid	485	0
Clay - very soft	977	2
Clay - soft	1947	4
Clay - fairly stiff	4914	6
Clay - very stiff	9772	12
Sand - dry	0	34
Sand - muddy	1947	30
Sand and gravel - dense	4851	34

To illustrate cohesion, think of two pieces of flypaper with their sticky sides in contact. Considerable force is required to slide one sheet over the other, even if no pressure is applied to the sheets. The resistance to shear here is due to cohesion between the sticky surfaces.

The bearing capacity of cohesive soils, like clay, tends to be greatly reduced when excess moisture is present. It is the usual practice to add a granular material, sand, gravel or both. When gravel or crushed stone are used they tend to sink into the soft clay and disappear. The major reason for this is that gravel and crushed stone have a small surface area to weight ratio. It has been found, for example, that 1 g of small gravel has only 1,5 cm^2 of surface area while 1 g of medium to fine sand has approximately 125 cm^2 and therefore it only takes about 1/10 of the bearing capacity to support the same weight of sand as it does of gravel and therefore sand is better able to "float" on soft clay than is the gravel. Sand however lacks interlocking properties and it is probably best to lay a layer of sand over clay, with a larger gravel on top as wearing surface (Piette, 1952).

7.4.2 FRICTION. In contrast to the flypaper example given to illustrate cohesiveness, friction may be illustrated by placing two sheets of sandpaper with their sanded surfaces in contact. They may be slid one over the other when no normal force is acting on them. When a normal force is applied the resistance to sliding, or the shear strength, increases in direct proportion to the normal force (Spangler, 1951).

7.4.3 MOISTURE CONTENT OF SOIL. Moisture relationships in soils are referred to as consistency limits (liquid limit, shrinkage limit, plastic limit and plasticity index) which constitute an empirical measure of the physico-chemical properties of the colloidal fractions of the soil. Consistency limits can be readily understood if an attempt is made to visualize the various states through which a mixture of soil and water pass as the water content is reduced by evaporation. These states may range from a true liquid at very high moisture contents, a viscous liquid, a plastic liquid, a semisolid, and finally, in the case of cohesive soils, a solid.

As water is removed from a soil-water mixture the liquid loses its liquid properties and becomes viscous. As the moisture content is further reduced it will become more and more viscous until it merges into the state of a plastic solid. The moisture content at which this change occurs, that is, the minimum moisture content at which the mixture acts as a viscous liquid, is called the liquid limit. At moisture contents below the plastic limit the soil-water mixture becomes semisolid in character and its bearing capacity increases very markedly.

Finally, although the pore spaces in the soil have become very small and the inward capillary pressures are very high, the soil solids can no longer be forced together by these forces and the mass reaches its minimum volume in the drying process. The moisture content of the soil at which a reduction in volume reaches its minimum is called the shrinkage limit. Soils containing small quantities of colloidal-sized material may be weak and friable in this state while those with a high colloid content may develop very high strength and hardness at low moisture contents.

The plasticity index is the numerical difference between the liquid limit and the plastic limit. It represents the range within which a soil exhibits the properties of a plastic solid and is an indicator of the cohesiveness and general workability of the material. A soil with a high plastic index is usually tough, sticky and muddy in wet weather. A soil with a low plastic index tends to be friable, to dry rapidly, and to be easier to compact.

Moisture plays a large part in trafficability of soils besides affecting soil strength. It directly affects two other soil properties, slipperiness and stickiness. These two properties are critical when they occur in soils of low-bearing capacity. Slipperiness is the condition that exists when the tractive capacity of a thin surface layer is so small that wheels or tracks merely spin without giving the vehicle a forward motion. Stickiness or the property of adhesion adds greatly to the rolling resistance of a vehicle and also tends to make steering difficult.

7.5 BEARING CAPACITY OF SOILS

The bearing capacity of a soil is taken to be the ultimate loading per unit of contact area between an imposed load, such as a vehicle's tire or track, and the soil beneath it, which will produce a shear failure within the soil mass.

Initially the concept of flotation for a vehicle crossing weak soil was borrowed from hydrostatics and was considered somewhat analogous to buoyancy. This however is not the case because of the unknowns involved. For a tracked vehicle which presents a rectangular footing the following equation gives the safe load for two tracks:

$$W = 2 \cdot l \cdot b \left(c \cdot N_C + \frac{1}{2} \gamma\, b \cdot N_r\right) \quad \text{in which}$$

W = safe load, in kg
b = track width, m
l = track length, m
c = cohesion of soil (from Table 18), kg/m²
γ = weight of soil, kg/m³
N_C = pure number dependent upon soil friction angle, Ø (Figure 49)
N_r = pure number dependent upon soil friction angle, Ø (Figure 49)

If the area is oval, such as produced by a tire, then the equation will become:

$$W = n\, \pi\, r^2 \,(1{,}3 \cdot c \cdot N_C + 0{,}6\, \gamma\, r \cdot N_r) \quad \text{in which}$$

W = safe load, in kg
n = number of wheels
r = radius of circular bearing area, m
c = cohesion of soil (from Table 18), kg/m²
γ = weight of soil, kg/m³
N_C = pure number dependent upon soil friction angle, Ø (Figure 49)
N_r = pure number dependent upon soil friction angle, Ø (Figure 49)

7.6 RELATIONSHIP OF RUBBER-TIRED VEHICLE TO SOIL

The use of rubber-tired vehicles for off-road operations under conditions where formerly only track type units were used is becoming universal. Through improved design of vehicles and particularly of their tires, it is possible now to use these vehicles where formerly they were unsuccessful. A good deal of knowledge on the factors which influence the performance of a vehicle under off-road conditions has been accumulated but there is much yet to be learned.

7.6.1 TIRE CONTACT AREA. The area of contact between a tire and soil may be found from the following:

$$A = \frac{0{,}90 \cdot W}{A_p}$$ in which

A = area of contact between tire and soil
W = weight upon tire
A_p = air pressure in tire

The equation assumes that the tire sidewalls carry 10 percent of the load on the tire. Regardless of the loading on the vehicle, for a given air pressure in a tire, the ground pressure per m² is constant. Additional loading on the tire results in greater deflection of the sidewalls and an increase in the contact area, to maintain the constant ground pressure.

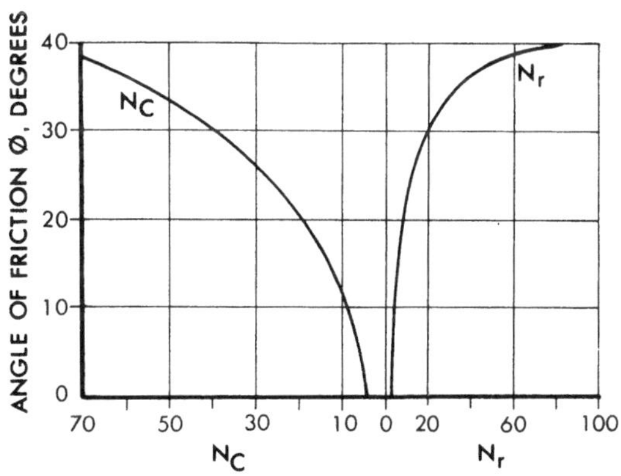

Figure 49. Soil parameters which affect the safe load of a vehicle.

7.6.2 RELATION BETWEEN TIRE AND SOIL UNDER STATIC CONDITIONS. Tire capacities are laid down in various tire manuals as to the safe allowable load the different sizes and types of tires will carry at stated tire pressures.

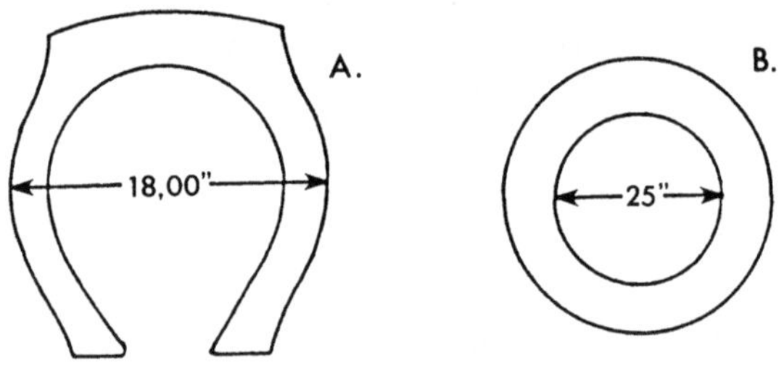

Figure 50. A. Cross section of tire indicating overall width at widest point, in example 18 inches. B. Side view of tire showing diameter of rim, 25 inches. Such a tire is referred to as an 18,00 x 25 size tire or the equivalent in centimeters (Anon, 1961).

Table 19 is an example of a table which relates service conditions, tire loading and tire inflation pressure (Goodyear, 1970).

Tire sidewalls do carry some of the imposed load although the entrapped air in the tire carries most of it. At a given loading a tire will deflect until sufficient area is in contact with the ground for the pressure under the tire to be equal to the pressure within it. To allow for the loading taken up by the tire sidewalls, the reaction at the surface of the soil may be taken as 110 percent of the tire pressure. For example, a tire with an air pressure of 82 kPa will exert $\frac{110}{100} \times 82 = 90{,}2$ kPa force on the soil.

For static loading, then the ground-bearing strength must be 110 percent of the tire pressure for a condition of no penetration. If the tire loading exceeds the ground-bearing capacity, the wheel will sink until it reaches solid footing or until the tires have gained additional contact area, sufficient to support the tire.

Table 19. Relation between service conditions, tire loading and tire inflation pressure (after Goodyear, 1970)

Service Conditions	Suggested Inflation
Firm Flat Ground	
Skidder	110 - 124 kPa
Forwarder	165 - 193 kPa
Hilly Firm Ground	
Skidder	96 - 110 kPa
Forwarder	138 - 165 kPa
Muskeg or Swamp	
Skidder	82 kPa
Forwarder	124 kPa

Tire Size	Ply	Rating	Load in Kilograms 1500	1750	2000	2250	2500	2750	3000	3250	3500	3750	4000
			Tire Inflation - kPa										
16,9-30	6	10	87	113	153								
18,4-26	6	10	82	107	122	140	165						
18,4-30	6	10	78	93	109	126	148	178					
18,4-34	6	10	65	79	95	103	121	131	161	183			
18,4-38	6	10	50	67	84	103	121	131	161	183	202		
23,1-26		10		41	55	68	82	97	113	129	144		
23,1-30		10			40	54	68	82	99	112	132	147	
23,1-34		10					68	78	90	103	117	132	148
24,5-32	10	12						70	78	85	95	106	118
30,5-32	10	12								64	71	79	87
67x34,00x25		10		24	32	40	47	55	64	72	81	90	100
67x34,00x30		10		26	36	47	55	64	73	83	94	105	117

To choose the most suitable tire:

1. Select service condition
2. Select suggested inflation pressure for that service condition i.e. swamp, skidder = 82 kPa
3. Calculate individual wheel loads at maximum loaded G.V.W. i.e. 2500 kg. Follow down 2500 kg load column until chosen inflation pressure is reached = 82 kPa
4. Suggested tire size is in left hand column = 10 ply 23,1-26 tire.

7.6.3 RELATION BETWEEN SOIL AND THE TIRE IN MOTION. Traction may be defined as the ability of wheels (or tracks) to grip the ground without slipping. Traction is provided on one hand by engine power and weight on wheels and on the other hand by the soil's ability to resist shear. The action of a tire is that it tends to dig into the upper layer of the soil and to shear it. When a layer of weak soil overlays hard soil, some tires are not able to dig down through to the underlying hard soil and an apparent lubricating effect known as slipperiness develops. In a soft soil where the bearing capacity is low a vehicle will settle and it will be necessary for the vehicle to push soil out of the way in order to move. This increases the rolling resistance.

As torque is transmitted to a wheel there will be some slip between the soil and tire before stress builds up to support the thrust. Slip is defined as the percentage of wheel turning velocity that does not contribute to vehicle motion. As slip occurs tire and soil deform and generate rolling resistance. As torque, delivered to the wheel, overcomes rolling resistance the vehicle moves (Olsen and Gibbens, 1983).

Traction is usually measured by a coefficient which is taken as a percentage of the weight on the driving wheels of a wheeled-type tractor or as a percentage of the track-type tractor weight. For average earth conditions approximately 85 percent of the track-type tractors weight can be developed in drawbar pull. For wheeled vehicles approximately 55 percent of the weight on the driving wheels can be developed in rim pull. Rim pull is the measure of the vehicle's ability to do work.

Table 20. Coefficients of traction for various ground conditions (Caterpillar, 1976)

Soils	Track-Type Vehicles	Wheel-Type Vehicles
Firm Earth	0,60	0,55
Loose Earth	0,60	0,45
Loose Sand	0,30	0,20
Ice	0,27	0,12

For wheeled vehicle

$$\text{-coefficient of traction} = \frac{\text{rim pull available (kg)}}{\text{weight on drive wheels (kg)}}$$

and rim pull available (kg) = weight on drive wheels (kg) x coefficient of traction.

When considering track or wheeled equipment no fixed drawbar pull exists. If drawbar pull is considered to be the actual force at the drawbar which can accomplish useful work under a given set of conditions, it is dependent upon the engine power and the weight of the vehicle but, in the final analysis, it depends on the tractive conditions of the soil on which the vehicle is being operated.

If all the power that can be delivered to the drive wheels of a vehicle is connected to the pulling force, then the rim pull and the drawbar pull are equal and no more pulling force can be produced than has been delivered through the final drive to the tires. Rim pull can exceed drawbar pull but not vice versa. However, when rim pull exceeds drawbar pull the excess power delivered is not converted into useful work but serves only to spin the wheels.

Table 21. Drawbar pull (kg) required to overcome log skidding resistance - log lengths (Caterpillar, 1976)

Grade	Load Weight - 2000 kg
30% upgrade	2410
20% upgrade	2200
10% upgrade	1990
Level	1780
10% downgrade	1570
20% downgrade	1360
30% downgrade	1150

Note: $\% \text{ grade} = \frac{\text{vertical rise}}{\text{horizontal distance}} \times 10$

In summary then:

- drawbar effort or pull is a variable factor governed by ground tractive conditions
- drawbar pull may be equal to but cannot exceed possible rim pull
- actual drawbar pull at any time will be equal to the rim pull that is being used
- rim pull is dependent upon equipment design and for a given machine will have a maximum assigned value that can be developed in each gear range

To establish the rim pull which acts through the wheel of a vehicle to develop shear stress in a soil, the tractive effort may be expressed as:

$$F = \frac{T \cdot R \cdot E}{r}$$ in which

F = the available tractive force (rim pull) - kN
T = engine torque - Nm
E = efficiency of gears and other losses, expressed as a decimal
R = total gear ratio, engine to axle
r = loaded radius of tire, m

Table 22. Values (E) for losses due to friction in gears and axle

Element	Value
Direct Transmission	0,90
Reverse	0,80
Other Gear Speeds	0,85

The torque/rpm curve can be developed for any particular engine. For example an engine at 2000 rpm could have a torque of 230 Nm. A tire (11,25 x 24) has a rolling radius of 0,557 m. The total gear reduction in third gear is 18,04.

$$F = \frac{230 \cdot 18,04 \cdot 0,85}{0,557} = 6332 \text{ N} = 6,3 \text{ kN}$$

7.6.4 ROLLING RESISTANCE. Rolling resistance is equal to the pull required to move a vehicle on the level and includes the energy necessary to compact or displace the material over which it is travelling and to surmount irregularities of the surface. No one definite fixed rolling resistance for a particular type of surface exists because it varies with tire size, speed and inflation pressure.

High tractive resistance of mud is due in part to the necessity of squeezing mud away from the tire as it rolls through the soft surface layer, and in part to a certain springiness of the whole soil crust. When mud is deep and sticky a wheel will scarcely turn and tends to slide.

The effect of a spongy subgrade under a thin running surface, a condition commonly found on lightly travelled forest roads, is greatly increased tractive resistance. The yielding of the subgrade under load, even if the running surface remains unbroken, may add, up to approximately 50 percent, to the tractive resistance.

When a wheel rolls over an earth surface there is a distortion of that surface. If the surface is of a yielding or plastic nature a rut will be formed. If the surface is elastic it will deflect under load but resume its original position. In the former case considerable rolling resistance is encountered, in the latter very little.

When a tire is in motion the total loading in the ground beneath it is greater than the static pressure due to weight and tire air pressure. Rolling impact pressures, which may amount to 30-50 percent of the static ground-bearing pressure value must be added to it.

Figure 51 shows the rolling resistance of a wheel on soft ground. The equation established can be applied to determine the relative ease with which different size tires can roll over rough ground or to show that, for rough ground, large wheels will require considerably less effort to overcome rolling resistance than small diameter tires.

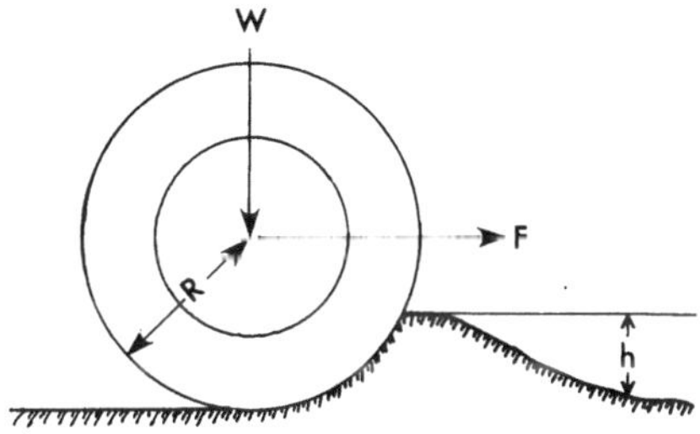

Figure 51. Rolling resistance encountered by wheel on soft ground.

F = force required to overcome rolling resistance - Nm
W = load on wheel - kN
R = rolling radius of wheel under load - m
h = height of obstruction - m

$$F = \frac{W \sqrt{2 \cdot R \cdot h - h^2}}{R - h}$$

Using the above equation a comparison is made of the forces required to overcome the resistance offered by an obstruction 15 cm high. An axle loading of 49 kN is assumed. With large single wheels the load on each wheel will be 24,5 kN. With smaller dual wheels (four wheels per axle) the load on each wheel will be 12,25 kN. With the smaller dual wheels the result for one wheel should be doubled to be comparable to the rolling resistance of a single big wheel. Rolling radius for the smaller tires (9,00 x 20) is 0,456 m, for the larger single wheels (11,25 x 24) is 0,557 m.

For the small wheels

$$F = \frac{W \sqrt{2 \cdot R \cdot h - h^2}}{R - h} = \frac{12{,}25\sqrt{2 \cdot 0{,}465 \cdot 0{,}2 - 0{,}2^2}}{0{,}465 - 0{,}2} = 17{,}7 \text{ kN}$$

Rolling resistance for the smaller diameter wheel should be doubled as it is one of a set of dual tires, to give a rolling resistance of 35 kN.

For the larger wheels

$$F = \frac{W \sqrt{2 \cdot R \cdot h - h^2}}{R - h} = \frac{24{,}5\sqrt{2 \cdot 0{,}5575 \cdot 0{,}2 - 0{,}2^2}}{0{,}5575 - 0{,}2} = 24{,}7 \text{ kN}$$

The dual wheels show a considerably greater rolling resistance (30 percent) than the larger diameter single wheel.

7.6.5 TIRES VERSUS TRACKS. Rubber tires have proven to be more effective on skidders and forwarders than tracks even though tracks perform well over a wider range of soil conditions. With tracks the lower ground pressures and the utilization of the entire weight of the vehicle in tractive effort results in good cross country performance. The discontinuity of ground contact on a wheeled skidder is a drawback compared to a track.

The use of tracks on skidders and forwarders has presented substantial problems due to the fact that the machines are operated under loaded and unloaded conditions and track tension under these widely varying conditions varies from being too great, with high friction losses or too loose so that the tracks come off. These conditions are in contrast to army vehicles or tree harvesters or processors where the weight on the tracks is relatively constant.

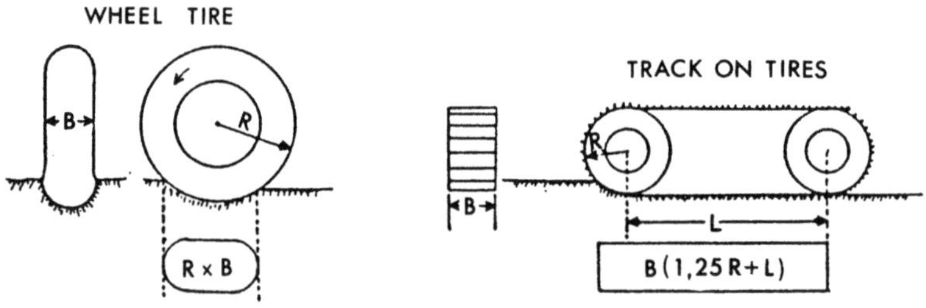

Figure 52. Tires versus tracks

Skidders and forwarders normally operate on trails which have been prepared to a greater or lesser degree (Figure 53, 54, 55). Tree harvesters and machines used in site preparation and regeneration are called upon to traverse cutover forest areas, to completely cover the area regardless of obstacles and ground conditions. Under such conditions tracks should prove more efficient than wheels in obtaining a total coverage of an area.

Figure 56 shows the drawbar pull for a crawler tractor and the rim pull for two sizes of wheeled skidders. From it, it is possible to select the different capabilities of the units at different speeds. For example, at 4 km/h in 3rd gear the crawler tractor (104 kW) shows a drawbar pull of 5500 kg, the 130 kW wheeled skidder 3500 kg rim pull and the 89 kW skidder 3000 kg rim pull.

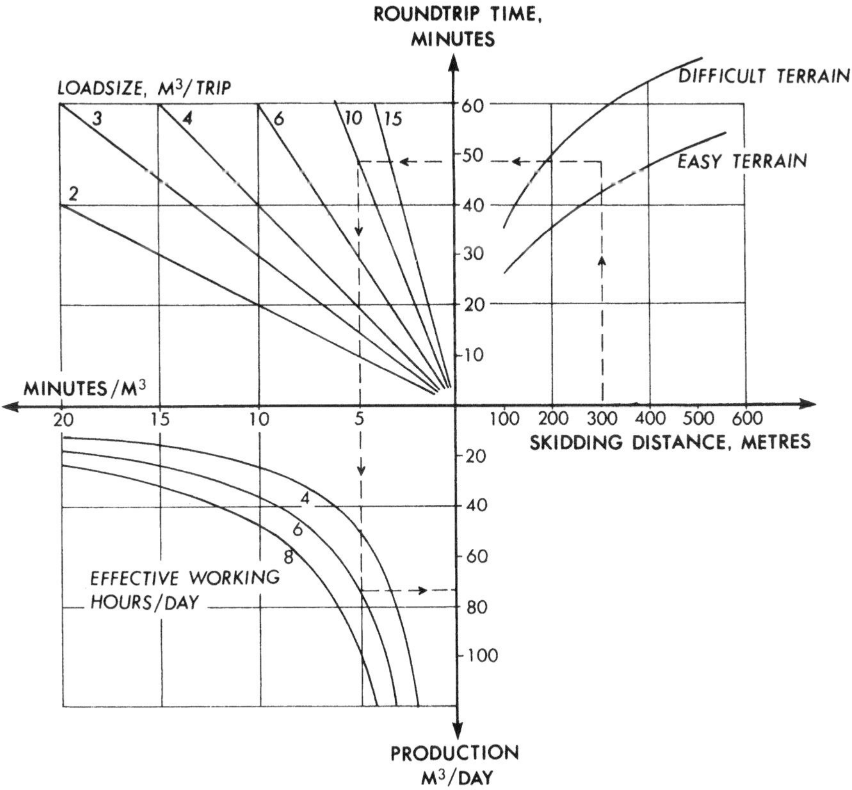

Figure 53. Skidding with crawler tractor (Bendz and Järvholm, 1970).

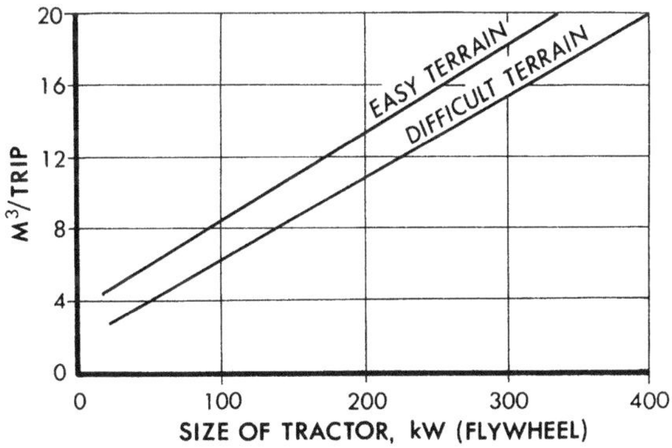

Figure 54. Typical load sizes with crawler tractors (Bendz and Järvholm, 1970).

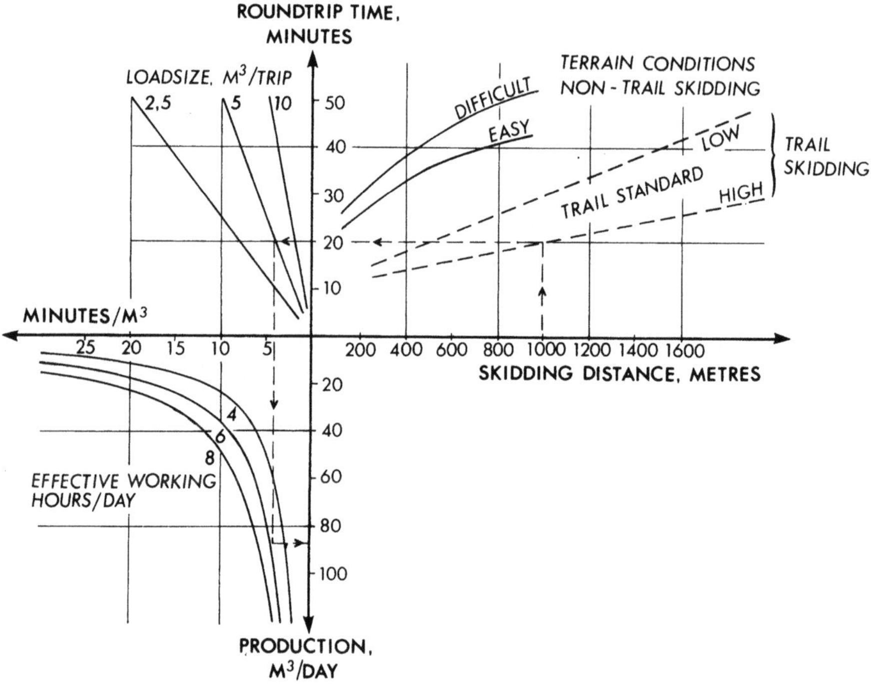

Figure 55. Skidding productivity with wheeled skidder. Load size will depend upon log size but will not exceed 10-12 m³ in most cases.

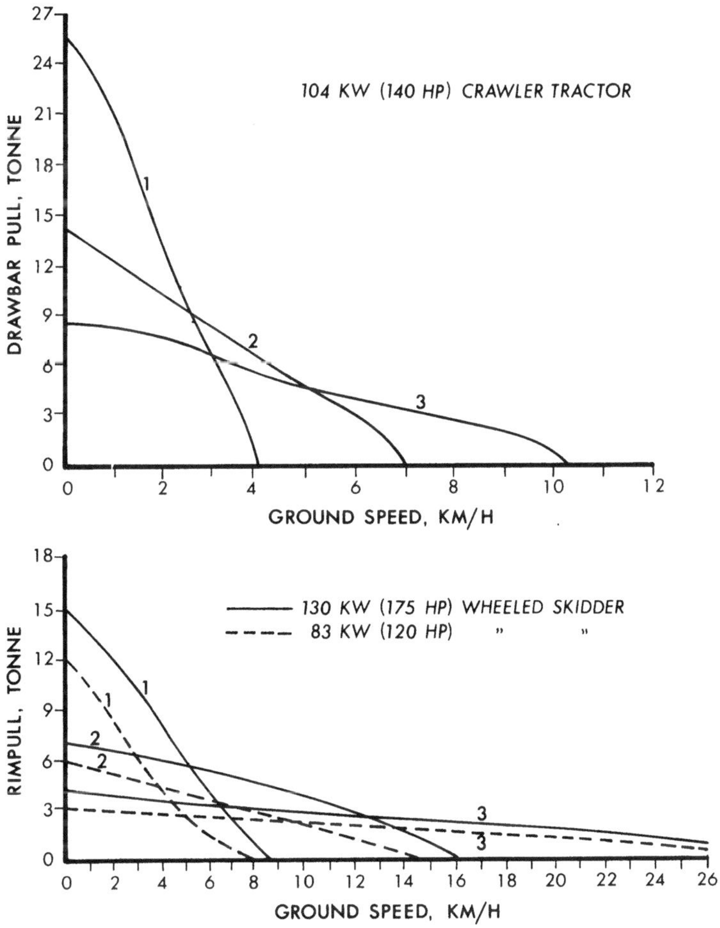

Figure 56. Comparison of crawler tractor and wheeled skidder.

8. TERRAIN

A general description of terrain (the physical features of a tract of land) in relation to the passage of vehicles is given by Nataraj (1969). It is based more upon description than an attempt to quantify the various elements statistically. Four categories were established:

Good - a terrain which permits most vehicles (wheeled and tracked) to move across its surface in any direction, at maximum possible ground speed, with the least amount of interruptions by obstacles or micro-relief.

Fair - a terrain in which cross-country movement is possible in many directions, but not to the same degree as defined for good trafficability conditions. The ground speed is interrupted by obstacles but they are not decisive.

Restricted - a terrain in which cross-country movement is restricted to few directions. The ground speed is slow and often is interrupted by obstacles. The path length tends to be tortuous.

Impassable - a terrain in which cross-country movement is extremely difficult or impossible, unless vehicles move on existing roads or tracks, and possibly along certain narrow valleys. Such areas contain few if any roads and are often tortuous. The ground speed is extremely slow.

Nataraj (1969), using the same four categories as outlined developed a table which adds more detail to his description. This is shown in Table 23.

Table 23. Key for trafficability (Nataraj, 1969)

Scale value	Main slope classification frequency of slopes greater than 60% (++)	Tree Density (Crown density in %)	Shrub density	Hydrographic features	Overall trafficability
A	0-5% (Plain or gently sloping) (++) Common	Rare or absent. No effect (0-15%)	Completely absent.	Surface water absent.	Good
B	5-15% (Rolling) (++) Rare	Patches Rows. Path slightly extended (15-25%)	Rare or absent - path selection easy	Seasonal flooding.	Fair
C	15-30% (++) Common	Moderately dense. Passable but length substantially extended (25-45%)	Moderately dense. Visibility somewhat affected. Path selection inhibited.	Small ponds easily by passed Marshy areas.	Restricted
D	More than 30% (++) Numerous	Dense. Passable but path tortuous (45-85%)	Dense. Visibility affected. Path selection difficult.	Unfordable water.	Impassable

9. TOPOGRAPHY

Topography together with soil determines to a large extent the general logging technique which can be economically used in an area. Slope is perhaps the most critical of terrain features to impact upon off-road transport or mobility of equipment. Combined with slope, ground roughness has proven critical.

9.1 SLOPE

Slope is expressed in either percentage or degrees. Table 24 shows various slope classes. Different classifications of slope have been developed and the following is representative (Löffler, 1984).

Table 24. Slope classes and gradient values (Löffler, 1984)

Slope Class	Gradient		Designation
	Percent	Degrees	
1	0 - 10	0 - 6	Level
2	10 - 20	6 - 11	Gentle
3	20 - 33	11 - 18	Moderate
4	33 - 50	18 - 27	Steep
5	50 -	27 -	Very steep

The three major characteristics of slope to be considered are gradient, shape or relief and length.

9.1.1 GRADIENT. Gradient in large degree controls the choice of logging system to be used. Ground skidding becomes more difficult and less productive with increasing gradient until it can no longer be used and is replaced by some form of cable skidding. The break-off point will depend upon the microterrain, the roughness of the slope surface and the gradient (Figure 58).

9.1.2 RELIEF. Slope shape or relief also impacts upon logging systems used and off-road mobility. Slope relief may be regular, undulating, terraced, concave or convex as shown in Figure 57.

A regular slope lends itself to ground skidding, particularly if the microterrain is not too rough. It is not a favourable profile for cable systems because of the cable deflection between supports. An undulating slope will not exclude ground movement of vehicles although it may slow them down. Much depends on the micro-gradients encountered in the undulations. Terraced slopes permit movement along the terraces or contours and should also permit the use of cable systems with intermediate supports. A concave slope is the preferred shape for cable systems as it permits ample ground clearance or deflection of the cable between supports at top and bottom. The convex system favours ground skidding over cables unless intermediate supports are usable so that the cable with load can clear the ground when tangent to the curvature of the slope.

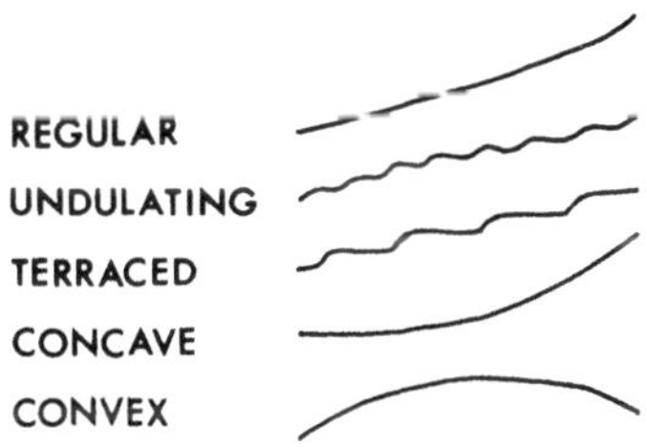

Figure 57. Slope shape or relief.

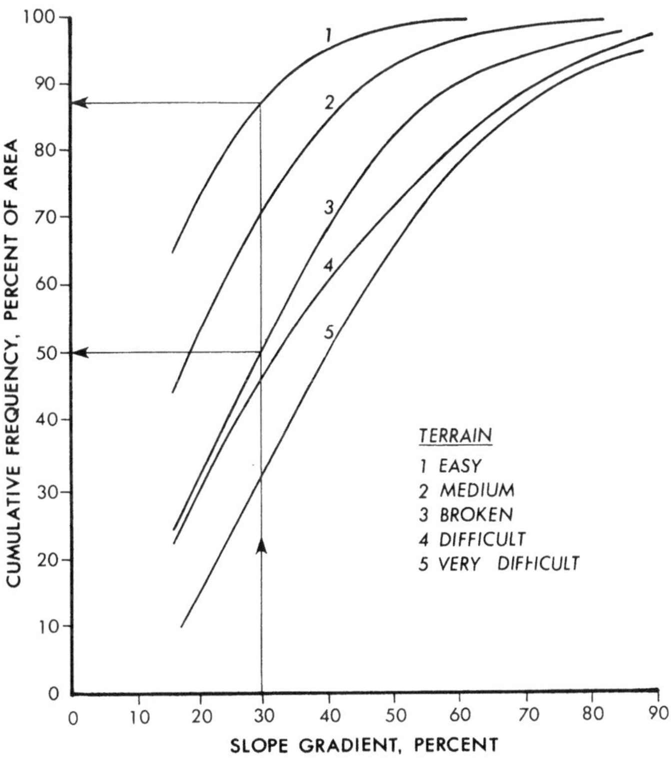

Figure 58. Classification of terrain difficulty according to frequency of slope gradient (Löffler, 1984). Example: With easy terrain approximately 90% of the surface area has a gradient of approximately 30% or less, with difficult terrain 50% of the surface area has a gradient greater than about 30 %.

9.1.3 LENGTH. The length of slope has some impact upon the choice of logging system. If slopes extend more than 300 m roads may have to be constructed along the contour of the slope. It is often possible to operate wheeled skidders on slopes up to 80 percent as long as great care is observed. The major problem encountered by the wheeled skidders is to reach the felling area. This is normally accomplished by constructing a return route which may be four or five times as long as that of the loaded skidder. When the load is assembled by the skidder in the normal manner the operator then drives the machine straight down the slope. Additional braking over that provided by the skidder may be necessary and this can be achieved by allowing the tree butts as well as the tops to drag on the ground. If the operation takes place in snow further braking is obtained by lowering the skidder's dozer blade (Bjerkelund, 1971). Trees are best felled along the contour of the slope and away from standing trees in order to facilitate delimbing and choking for transport by skidder. This work can be heavy and dangerous and often bonuses are paid as compensation.

9.2 GROUND ROUGHNESS

Ground roughness as it affects off-road transport is determined by the size (height or depth), and frequency of occurrence of obstacles. Löffler (1984) classifies ground roughness based upon these two characteristics in Table 25 and in Figure 58. Spacing of obstacles is a measure of frequency of occurrence.

Table 25. Classification of ground roughness (Löffler, 1984)

Height of obstacle (cm)	H-20 10-30	H-40 30-50	H-60 50-70	H-80 70	Ground roughness class	Description
Occurrence	Infrequent	Isolated	-	-		
Number/ha	40-400	<40	-	-	1	
Spacing/m	5-16	>16	-	-		
Occurrence	Moderately frequent	Infrequent	Isolated		2	Slightly uneven
Number/ha	400-4000	40-4000	<40			
Spacing/m	1,6-5	5-16	>16			
Occurrence	Frequent	Moderately frequent	Infrequent	Isolated	3	Uneven
Number/ha	400-4000	400-4000	400-4000	<40		
Spacing/m	1,6	1,6-5	5-16	>16		
Occurence	Moderately frequent	Moderately frequent	Moderately frequent	Infrequent		
Number/ha	400-4000	400-4000	400-4000	40-400		
Spacing/m	1,6-5	1,6-5	1,6-5	5-16	4	Rough
Occurrence	Frequent	Frequent	Infrequent	Infrequent		
Number/ha	>4000	>4000	40-400	40-400		
Spacing/m	<1,6	<1,6	5-16	5-16		
All surfaces with ground roughness more tham that of Class 4					5	Very rough

10. SOIL COMPACTION

With the introduction of heavy machines on timber harvesting operations came the realization of possible damage to forest soils. The surface layer of forest soils are typically porous well-aggregated mixtures of mineral soil particles and organic material and it is this porous nature that makes the forest soil particularly susceptible to compaction by logging equipment. A single pass or very few passes will make the greatest impact on the bulk density of these porous soils. Disturbance of the upper surface layer along with its compaction has produced long-term damage. The compaction process has been clearly described by Li (1956) "If applied stresses exceed the shearing strength or resistance of the soil, local failure begins and the load begins to sink into the soil. As the load sinks, the soil under the load is pushed downward and outwards. This motion mobilizes more and more resistance, consisting not only of the increased resistance due to lateral confinement from depth but also that due to the increase in soil density that results from the settlement motion itself, provided that the soil is not completely saturated. The settlement stops when equilibrium between stresses and resistance is reached."

When the load sinks into the soil it replaces the volume of soil by pushing it in an outward direction. This action may result in compaction of the soil under the load and the loosening of the soil on the sides.

Soil compaction is most critical in those regions which do not experience soil freezing and it has been found that soils in these regions may take up to 50 years to recover. The freezing of wet soils results in the formation of ice. The attendant 9 percent increase in the volume of moisture may result in a reduction in soil density. With the redistribution of the thaw water in the soil there can be a regain in soil density. During the frost-melting period a reduction in soil strength or bearing capacity is encountered. The cycle of freezing and thawing, where it occurs, tends to reduce the effect of compaction. The moistening and drying of clay particles with resultant swelling and shrinking also tend to loosen the soil structure.

The compaction process is a reduction in volume for a given mass of soil, expressed as an increase in the bulk density of the soil. Bulk density is the oven-dry weight of a soil expressed on a volume basis (Armson, 1979). For example, if the oven-dry weight of a soil sample is 140 g in a container of 100 cm^3, the bulk density is 140/100 = 1,40 g/cm^3. Armson states that in general there is a progressive increase in bulk density, from the surface of the soil downwards. This gradient in bulk density reflects the fact that organic matter in the soil is to be found mainly in the upper layers of forest soils and the bulk density of such organic matter is much less than that for mineral soils.

Compaction alters the soil structure which determines the water movement and content. The soil particles are pressed together into a closer state of contact and as a result the contained air and/or water is expelled from the soil mass. Water content is most affected by changes in the volume of the voids or spaces between soil particles. Compaction occurs most severely with soils that have different sizes of particles as this allows the voids between the larger particles to be filled by the smaller particles. For example, spherical particles of the same diameter, with tetrahedral packing will result in 74 percent solids and 26 percent pore space. If particles of much smaller diameter are added, the solid volume can be increased to 93 percent of the total (Lutz, 1952).

Compaction usually produces layers of soil with high bulk density rather than a uniformly compacted soil. Such layers, called hardpans, are most commonly found in medium-textured soils, loams, sandy loams and silt loams.

Studies indicate that the bulk density of soils increases rapidly with the first few passes of a vehicle and then the rate of change in soil density decreases with successive trips. Lenhart (1986) found that bulk density reached a maximum after four trips and remained statistically constant with additional trips. Froelich et al (1980) found that in a study of 20 trips over a skid trail, after the sixth trip the increase in soil density was very slight. The increase in bulk density as a result of vehicle traffic has been found to be negligible below 30 cm from the surface.

Figure 59 illustrates the effect that increases in soil density have upon the growth of seedlings.

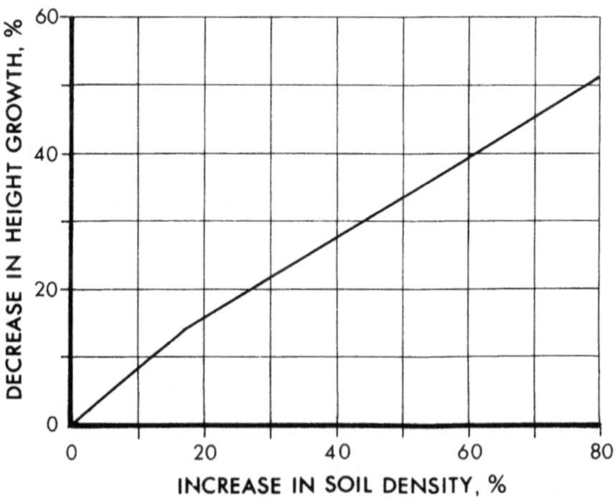

Figure 59. Relationship of increase in soil density to seedling height growth (Rollerson and Kennedy, 1984).

The wide variation of soil and its moisture content affect soil-vehicle equations to the point where their application in a working environment can be questioned. However, as described above, the impact of vehicles over forest soils and the mechanisms of soil compaction have been established. With increased mechanization in the forest soils, regeneration and the residual stand are coming under increasing study.

11. EARLY THINNINGS

11.1 HARVESTING SMALL WOOD

To obtain high growth in a juvenile forest stand more trees are needed per unit area than what can survive in later periods of the rotation. Close spacing between the trees - up to a point - improves the quality of wood and gives, in thinnings, a wider choice in the selection of trees and species to survive for sustained growth. Also natural regeneration of trees may come into a plantation in which a prescribed spacing is observed. These factors alone or combined create a situation in forest management when it is desirable to carry out early thinnings, the purpose of which is to regulate the stand in spacing and composition for optimal future growth and also to capture material that would be lost to natural mortality.

However, let us first dispose of the problem of small wood in another setting - in a regular harvesting operation. Here the criterion for its utilization is that the revenues from the minimum-sized piece of wood shall exceed the variable cost for harvesting it. Some qualification can be tied to and slightly modify this criterion. But in exploitation forestry, as explained in Part I, (p. 10) of this book, the criteria are somewhat different.

Returning to the thinning operation it is obvious that there is a combined silvicultural and logging problem. It is most accentuated for the first thinning operation during the rotation of the stand for which the following text applies.

The combined problem of the thinning operation is to maximize the value of:

Revenues - Costs + Value of the residual stand

This formula applies for decisions regarding treatment of single stands, but the same general principle rules in studies of optimal silvicultural regimes, e.g. prescribing number and timing of thinning operations. We shall not concern ourselves with this management problem, but as an example focus our attention on how tree and log size affect harvesting in managed forests in Northern Europe. Where trees grow to larger sizes such as in the radiata forests in subtropical regions or where markets are depressed the same general problem is encountered even if critical tree sizes for satisfactory economic return of the thinning may be considerably larger.

Non- or low-mechanized operations, even if outdated in most industrialized countries, will be included as they still should be applicable in developing countries with low labour wages.

It has already been pointed out (Figure 1.8, p. 35 in Part I) that what is a small tree should be seen relative to the normal tree size for which the technique is designed. A small tree in tropical logging is often of the same size or larger than the average tree in the final harvest of a coniferous stand. The following will be an extension of the analysis of the influence of tree and stand factors presented in Chapter 6, Part II.

Two work elements in particular are influenced by tree size, the felling and the manual handling of wood. Figure 60 shows the labour input in the felling and conversion (limbing and cross-cutting) of very small trees (Sundberg, 1954). This graph can be used to determine the smallest tree or log for which the variable costs and the revenues will break even.

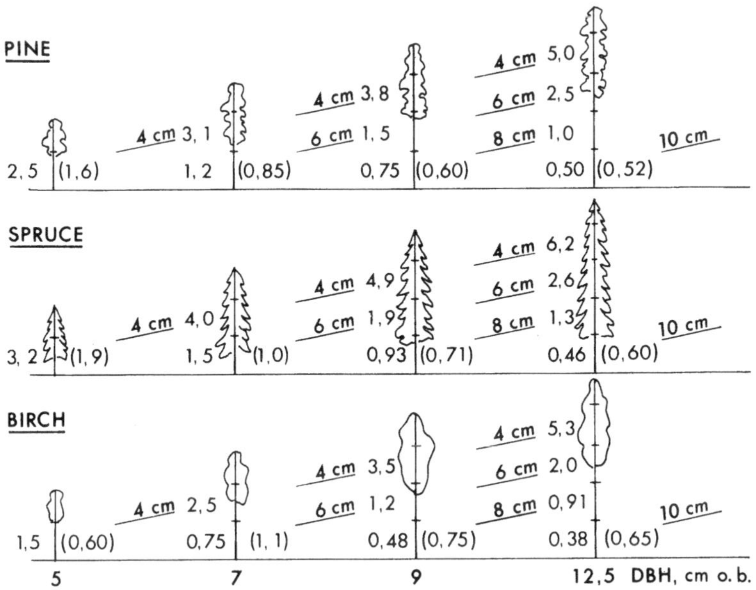

Figure 60. Example of the relative time consumption for the felling and conversion of small trees of pine (P. sylvestris, L.), spruce (Picea abies, (L.) Kart.) and birch (Betula sp.) for tree sizes yielding 1, 2, 3, and 4 logs of 3 m length with the corresponding mid-diameters 4, 6, 8, and 10 cm. The ratio 1 approximately equals 0,17 man-days (md) per m^3 solid wood when using powersaw and 0,20 md/m^3 using bowsaw and axe. "Tree time" (walking to tree, felling etc.) is allocated on the butt log and shown in brackets in man-day per m^3 for that log. Example: A pine tree of DBH 7 cm will require $0{,}17 \cdot 3{,}1 = 0{,}53$ md/m^3 for the top log and $0{,}17 \cdot (0{,}85+1{,}2) = 0{,}35$ md/m^3 for the butt log when using powersaw. The figure illustrates the crown length and branchiness of the studied stands (Sundberg, 1954, based on an unpublished study by The Forest Operations Institute, Stockholm).

The delimbing of small trees is very costly and varies much with the branchiness which can be very different for logs of the same size depending on the location of the log between butt and top. This can be exemplified by the numbers in Figure 60 for a pine log of 4 cm mid-diameter: 2,5 - 3,1 - 3,8 - 5,0 md/m^3 for butt to top log.

The influence of size is less accentuated in extraction and haulage. The possibility of reducing the impact of tree size in mechanized operations is described in detail in Chapter 6.2, e.g., by bundling or with a feller buncher with an accumulator head. However the advantage of such an accumulator is considerably decreased in selective thinnings as the felling head must be moved back and forth to reach the trees. Line or corridor thinning will eliminate these constraints but results in a silviculturally less satisfactory stand. As the delimbing of these small trees often accounts for 60-70% of the cutting time, full-tree harvesting of small trees can reduce the cost considerably. Such systems are described in Chapter 12.

11.2 MANUAL HANDLING OF WOOD

On many logging operations wood is manually handled, e.g. dragging or piling in the stump area by the cutter to facilitate the following transport. In countries with low labour wages especially, manual handling is the most economical procedure in cutting young stands with small tree dimensions. The following indications of the influence of log size on the manual handling are derived from a study made under standardized work conditions in an open field without any of the obstacles usually appearing in a forest (Sundberg, 1959). The production rate and work efficiency are therefore higher, and probably optimal, as compared to real work situations in the forest. However, the ratios for different log sizes should be approximately the same.

Figure 61 shows the production rate of a well-trained forest worker in horizontal kgm/min (curve A) and the physiological work efficiency in horizontal kgm/kcal for dragging logs a distance of 10 m (curve B). The return empty-handed to pick up the next log is included. Thus, the numbers apply to a complete work cycle. It can be seen that the optimum log weight is about 60 kg and that smaller log sizes result in rapidly decreasing production. Large log sizes give lower production and the decline is even more accentuated in physiological work efficiency than in production per unit time.

It should be observed that the "horizontal" kgm is production rate and not mechanical work. If we assume that, in dragging a log weighing 50 kg, 30% of the log weight is mechanical work in the form of friction between one log end and the ground (as measured in the tests), the efficiency of the worker to convert energy to mechanical work is:

$$(0,3 \cdot 114)/4,27 = 0,08 \text{ or } 8\%$$

(The number 114 read on the graph, 1 calorie = 427 kgm).

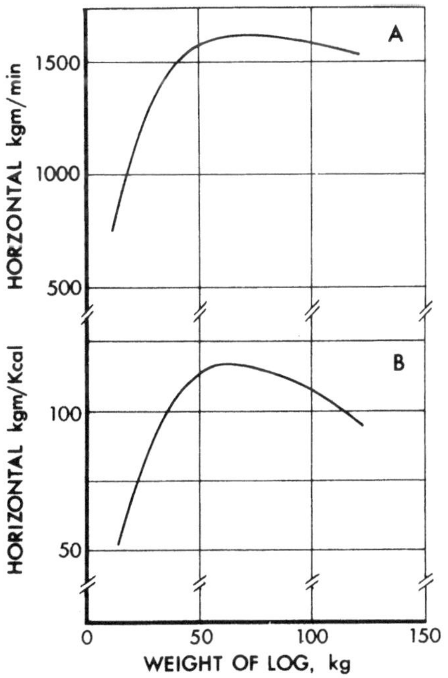

Figure 61. Production rate in horizontal kgm/min (A) and physiological work efficiency (B) in horizontal kgm/kcal (of the workers' energy input) for the manual dragging of logs of different weights. The number refers to a complete work cycle and thus includes the return empty to pick up the next log. (Results from tests under standardized conditions).

Manual piling of wood in stacks or steres is often carried out to form bundles or units for storage, drying and for loading. Figure 62 is an example of the influence of log weight on the production rate (graph A) and of the physiological efficiency in the work (graph B). The graphs derive from standardized tests (Sundberg, 1959) in which the logs are picked up by the worker from the ground, moved (a short measured distance) and piled at various heights. Both the production rate and the physiological work efficiency increase for increasing log weight except for a piling height of about 1 m, at which height logs exceeding about 40 kg in weight give both lower production and physiological efficiency.

The conclusions are that the optimum piece weight in the manual handling of logs is in the range of 20-40 kg and that piece weights over 20-25 kg may impose unduly high physiological work load on the worker and could contribute to health hazards for the workers. When the piece weights exceed this range some installation or equipment is definitely needed to facilitate the work and reduce the strain of the work for the worker. There are in the ergonomic literature many studies which support these recommendations. On the other hand, to retain good production in manual handling, tree boles should not be cut up unnecessarily into lengths giving piece weights lower than about 20 kg.

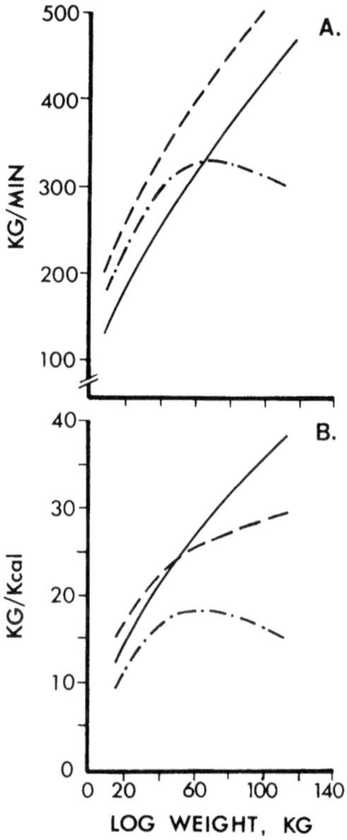

Figure 62. The influence of log size on the production rate (A) in kg/min and on the physiological efficiency (B) in kg/kcal in piling of wood (Results from tests under standardized conditions)

——— piling height 10-20 cm

----- " 70-80 cm

-.-. " 120-130 cm

11.3 HARVESTING WITH MINI-SKIDDERS

This recent development is a skidder on which the operator does not ride. The concept permits the downsizing of engine, hydraulics, wheels and tires. The costly cab for the operator is no longer needed and the mechanical engineering design is simple. The present generation of these machines was originally designed and built for hunters to bring large game animals out of the forest. The machines weigh 200-300 kg and their power rating is about 5 kW. They are constructed with wheels or tracks and have a maximum traction of about 4 to 5 kN. They have a low ground pressure, travel slowly at the speed of a walking man, and cause minimum if any damage to the residual stand. The turning radius when unloaded is less than 1,5 m and when loaded about 3 m.

Figure 63. Thinning with mini-skidder. One man with a powersaw and a skidder fells a tree over a roller mounted on the stake of the rear bunk to obtain a convenient work height for limbing and loading (Nordfjell, 1988).

Mini-skidders are being used increasingly in small-scale operations such as early thinnings in farm forestry but can also be used for salvage logging of windthrown scattered trees. Nordfjell (1988) reports that one man with a powersaw and a mini-skidder can produce 2,7 to 3,8 m^3 of pulpwood (1 m^3 approx. = 1 tonne) per 8 hour shift including cutting of thinnings of an average DBH of 11-13 cm and extraction to roadside, over a distance of about 100 m. The fuel consumption of the skidder is as low as about 0,4 l/tonne pulpwood. However, if the skidder is continuously employed in a team of fellers, its daily production will be considerably higher. Such use is relevant to developing countries, for which the low capital investment and the simple engineering design should be suitable.

In the 1960's and early 1970's mechanization was marked by successively bigger and more sophisticated logging machines aiming at higher labour productivity. Thinning to waste was looked upon by many as the final answer to the problem of small wood. However, escalating machine and fuel costs after the oil embargo in 1973-1974 and the renewed markets for fuelwood triggered a demand for new thinning techniques. The mini-skidders as well as the single grip harvester are examples of a response to these demands. Their use makes early and more frequent thinnings in silvicultural regimes possible and contributes to a change to more intensive silviculture.

11.4 THINNING WITH SINGLE GRIP HARVESTERS

The harvesters of the first generation were, as a rule, equipped with a felling head which, after severance of the tree from the stump, transferred the tree to a special mechanism mounted on the base machine. The tree was released from the grapple and in a second step seized by the feeding system of the processing unit. These machines, now often called double grip harvesters, collect most of the slash on or close to the strip road, which is not always desirable.

In the second generation of harvesters the processing functions (limbing and bucking) were moved to the felling head so that the whole cycle could be performed without releasing the grip on the tree, therefore single grip harvesters (Figure 64). The tree could at the same time be moved close to the strip road and processed logs be piled there for forwarding. The system allows a more even distribution of the slash over the area.

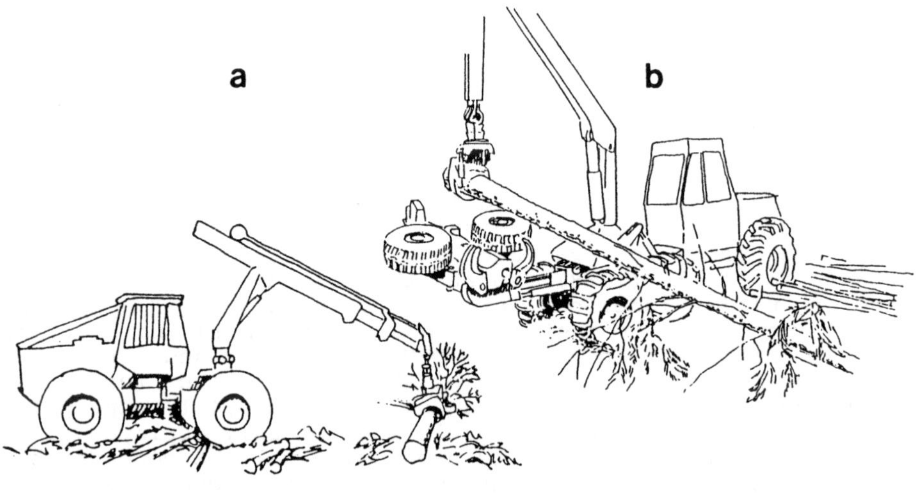

Figure 64. a) Single grip harvester which can fell and process a tree without releasing its grip upon it; b) Double grip harvester which transfers felled tree to processing mechanism mounted on the base machine (Courtesy: Forest Operations Institute, Stockholm).

In order to extend the spacing between strip roads a zone in between is manually felled with the tree tops towards the strip road so that they can be reached by the machine boom. The work procedure is further explained in Figure 65. Scherman (1987) reports a production of 60 to 90 trees per productive machine hour for a large machine, cutting diameters of about 50-55 cm. Smaller models with a maximum cut of up to 45 cm produce 80-120 trees per productive machine hour. Several makes of these harvesters are now available, for small trees as well as for tree sizes in later thinnings. The single grip harvesters have reduced the cycle time for the felling and processing of a tree and considerably improved the economics of early thinnings.

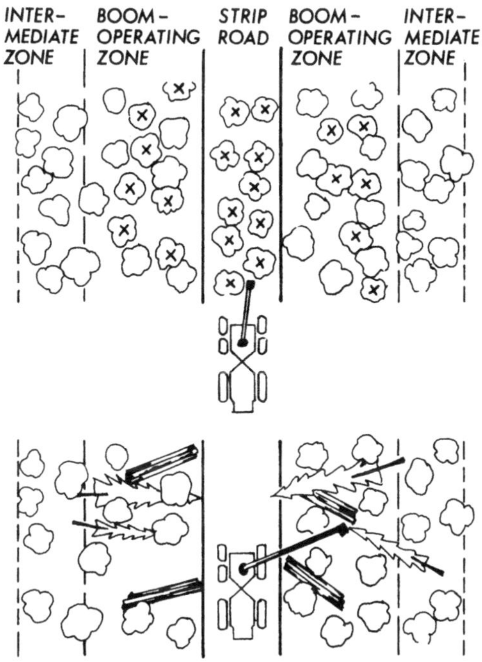

Figure 65. Thinning with a single grip harvester. The strip road is first clear-cut in one pass of the machine, at the same time selectively thinning the zone within reach of the boom. The intermediate zone is then manually felled, tops towards the strip road. In a second pass of the machine these trees are processed and logs piled along the strip road for forwarding to truck road (Scherman, 1987).

12. HARVESTING BIOMASS FOR FUEL

In the industrialized countries wood for fuel, after more than 100 years of declining use, is coming back as a market product. This has triggered the development of new harvesting techniques, each adapted to the special type of biomass resource available as fuel in that particular situation. A wide range of different types of woody materials are now used for very different market outlets, resulting in a corresponding flora of machines, systems and approaches. A comprehensive review is outside the scope of this text, in particular since the ongoing development is very rapid and the scene is continually changing. It will merely be attempted to present the principal impacts of the use of wood for energy or heat on the traditional logging techniques and related forest management problems. From the vast stream of publications reference is made only to a few, in which additional references can be found. For North America the subject is well covered in the proceedings of a series of conferences arranged by the Forest Products Research Society (FPRS 1982, 1983, 1984) and for Northern Europe, reference is made to Hakkila et al (1979, 1988).

Forest biomass for fuel from a logging point of view can be grouped as follows:

1. Full-tree utilization of forest for which demand from industry is low or non-existent - in United States it can be mixed hardwoods in the Lake States, in the Northeast and South and in Canada, in many of the provinces
2. Small trees and residues from logging operations of industrial wood
3. Energy plantations.

The economic viability of forest biomass for energy is strongly influenced by price fluctuations for other energy sources such as oil, gas and coal. However, it can safely be stated that wood fuel is now a market product on equal terms with sawlogs and pulpwood. It will stay and is likely to grow in importance as fossil, non-renewable energies become scarce and more costly. Therefore, forest management and logging should now be planned to integrate and exploit the new possibilities opened up by the energy market.

The new technologies for harvesting wood for fuel, with few exceptions, include two major features:

1. Harvesting of unlimbed trees or tree parts resulting in decreased logging costs because of elimination of costly delimbing, is of particular importance in logging of small trees. Also an increase of yield and revenues per tree and per unit area, can contribute to improved economics of earlier uneconomic operations.
2. Chipping at some point between the stump and the end use facilitates material handling, storage and auto-

matic feeding into burners. As wood chips are less bulky than unlimbed trees, truck haul is often preceded by chipping, in particular for distances over about 50 km.

Forest biomass for fuel in industrial applications has attracted more interest in Northern Europe and North America than elsewhere. A few examples will be described in the following paragraphs to illustrate modifications of machines and systems to meet and exploit the opportunities of energy for forestry. The Poor Man's Energy Crises in many developing countries have dramatically changed the attitudes of most aid agencies. They are now making major efforts to try to alleviate the extremely serious situation in many regions. The major problem now is to grow more biomass and protect the environment. There are no doubt technical problems which deserve more specific attention. However, they will not be included in this book.

12.1 SCANDINAVIAN EXPERIENCES

In Scandinavia, and elsewhere in Europe, all forests are used for sawlog and pulpwood production. Developments have therefore had an emphasis on woody material not usable for such purposes. A variety of equipment, for the forest farmers' own supply of energy from the forests or for nearby local markets, is available, such as portable chippers mounted on farm tractors. The review will be limited to more large-scale operations as these seem to be of more general interest.

A significant innovation is harvesting unlimbed tree parts in the first thinning of a stand. A comparison with the traditional thinning system, harvesting pulpwood from a forest stand with average DBH 10 cm, is given in Table 26 based on a report by Andersson and Björheden (1986). The two systems are illustrated in Figure 66.

Table 26. Comparison of traditional utilization of pulpwood and full-tree logging (unlimbed tree parts) in the thinning of a forest stand with DBH 10 cm o.b.

	Traditional system	Full tree logging
Labour input, man-day/ha	10,1	4,1
Yield: pulpwood (unbarked), m^3/ha	37	37
biomass for fuel, solid m^3/ha	0	29
Relative cost at roadside (all costs on the pulpwood)	1,0	1,02
Surplus, USD/ha, delimbed and debarked at millsite after 40 km truck haul	- 3	560

The estimated surplus in Table 26 refers to costs and prices in 1985. In this case the price for biomass for fuel is 85% of the pulpwood price. The estimate is based on the assumption that the unlimbed tree parts, usually cut up to length of 4-5 m, are delivered to a pulpmill with a drum debarker able to delimb and debark this material and also that the mill can make use of the fuel for its heat requirements or that there is a market for it close by.

The full-tree biomass can also be chipped at roadside, e.g., for delivery as fuel to a heat generating plant, an alternative giving somewhat higher surplus in the study than to a pulpmill. The chip is less bulky than a truckload of unlimbed wood, lowering the cost of the truckhaul. As the thinning preferably shall be made when the ground is free of snow and the demand of fuel peaks during the winter, the biomass should preferably be chipped at roadside before the winter. As this chip very quickly loses heat value from microbiological agents, long-term storage should be avoided. On the other hand, storage of unlimbed tree parts until the winter will involve chipping of snow-covered piles, also reducing its heat value. This is an example of the problem of organizing the material flow of wood fuel in an open energy market which also has its ups and downs because of the fluctuation of prices of oil and gas.

The drum-debarking of unlimbed tree parts should be made with a mix of ordinary pulpwood. The capacity of debarking pulpwood will decrease a little. Figure 66 shows that the total yield of pulp fiber will increase a little as compared to traditional thinning, in which smaller top sections of the bole are left in the forest.

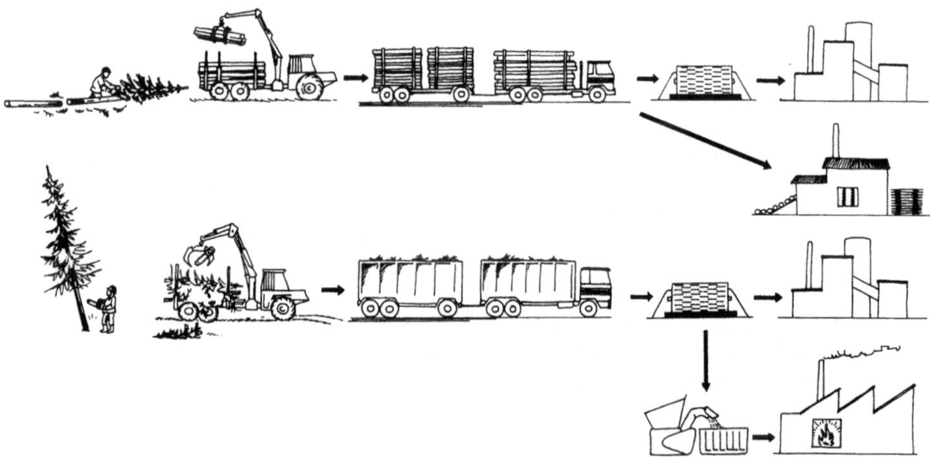

Figure 66. The two harvesting systems compared in Table 26.

This modification of full-tree logging in thinnings is very sensitive to even very small variations of the properties of the stand so the figures presented here should be regarded with caution. However, the method has greatly improved the economics of the early thinnings and contributed to the application of rational silvicultural regimes.

Figure 67. Forwarder loading unlimbed tree sections in the stand after bucking the trees with the powersaw mounted in the grapple.

Logging of residues in final fellings is now common. A tractor equipped with chipper and container can chip the residues in the logging area or at roadside.

In Scandinavia wood fuel is almost entirely used for production of heat. However, if the material is taken to a sulphate mill it can be fed into the energy system based on recovery by burning black liquor, producing both electricity and heat. In fact, modern pulpmills are net producers of energy delivering to an outside market both electricity and low grade heat for residential heating.

It is evident from the Scandinavian experience that wood fuel is being established as a regular component in forest management, in logging and also in the national energy systems. The end use can be of any scale from small operations by forest farmers for their own residential needs to large-scale applications by industry.

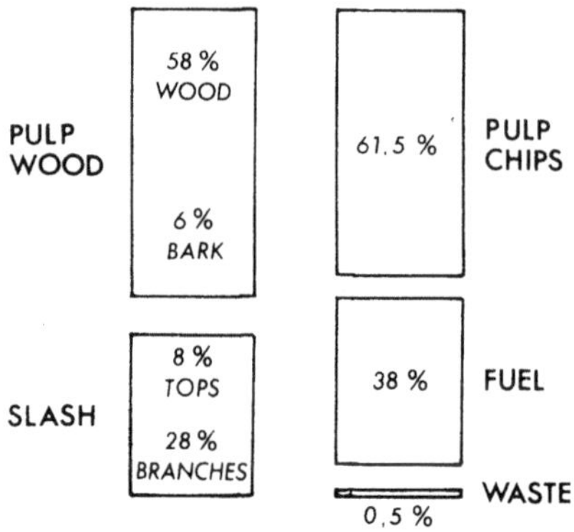

Figure 68. Utilization in thinnings, in percent of total biomass above felling cut.
Left: Traditional system - slash not utilized
Right: Modified full-tree system - unlimbed tree sections.

Figure 69. Forwarder collecting logging residues for transport to roadside.

12.2 NORTH AMERICAN DEVELOPMENTS

The consideration of energy as a legitimate forest product has not been widely accepted. Certainly 50% of the world's harvest of wood is used as fuel, predominantly in lesser developed countries, but the use of biomass instead of fossil fuel (coal, gas, oil) is still open to question and to economic analysis. Wood accounts for almost one half of the total biomass produced on earth, so that it is an important source of fixed carbon and energy (Whittaker and Woodwell, 1971). It may come as a surprise to many that in the United States the end use of most timber is not lumber or pulp and paper but fuel for energy and this energy use has been growing at a substantial rate (Koning and Skog, 1987).

The technology of large-scale combustion of biomass is well established and where it has been proven to be economically feasible is being successfully implemented. In some instances the wood-burning plants are stand-alone electric generating plants, in other instances thay are constructed to permit cogeneration of electric power along with process steam for an industry.

There has been an expressed reluctance on the part of Canadian and United States forest industries, with a few exceptions, to seriously consider the use of logging residues and presently unutilized trees and forest stands for energy purposes. The concern has been that if forest biomass is considered for energy its production and cost would impinge upon the production of pulpwood and other forest products.

One point should be made clear and that is that biomass produced for energy and that produced for pulping are not the same product. Pulp-quality chips could be used for energy purposes but energy-quality chips, almost without exception, cannot be used for pulping. A comparable distinction exists between sawtimber and pulpwood. Pulp chips have rigorous specifications as to size, species, cleanliness from bark and extraneous material, while energy chips have no specifications.

Wood and bark have an energy content of 18,5 MJ/kg on an ovendry basis to within 10%, regardless of species, density, quality, green moisture content, etc.

It is considered by some foresters that the use of biomass for energy could be the greatest boon yet encountered for forest management in North America. Because biomass for energy has no specification it will make it possible to low-grade our forests rather than to high-grade them. All the former green junk in the so-called silvicultural slums, the dead, dying, and down material would have a value for energy. Non-commercial thinnings might now break even or be profitable rather than be a cost. Energy plantations of fast-growing hardwoods are another practical and economic source of feedstock for a biomass-based power plant.

A major return to the community and to the country as a whole would be through the creation of a new product which did not exist before. The purchase of residual oil for combustion entails a direct payment to an agent outside the country which affects the country's balance of payments. The money paid for this oil provides no multiplier effect and, if anything, makes a negative contribution to the local economy, as compared with biomass. The money spent on biomass used for energy will have a multiplier effect similar to that generated by other conventional forest products although perhaps not so great as it does not involve a secondary manufacturing sector.

In Canada, an analysis of nineteen studies of logging residues found that the average mass of residual material following a logging operation is not clearly dependent upon the logging system, whether full tree, tree length, or shortwood and averages 0,21 ODt/m^3 harvested for conventional use. In mixed wood stands the residual trees form the largest part of the material whereas in pure softwood stands it will be the unmerchantable tree components which dominate.

Woody biomass may be removed from the forest prior to the harvest of trees for conventional products, it may be removed at the same time, as in whole-tree logging, or it may be collected after the regular harvest. Studies have shown that the collection of biomass is most cost effective with a single-pass operation, where the energy biomass is removed at the same time as the wood taken for conventional use.

The key to the efficient production of biomass for energy is similar to that for any forest operation. It is the attaining of a proper balance of equipment so that all phases of the harvesting process perform at a level which supports the productivity needs of the next phase and that no bottlenecks develop in the operation.

Generally biomass with diameters of less than 10 cm and/or concentrations of less than 100 t/ha will not be economically profitable using currently available equipment (Forestal, 1983).

In many instances conventional harvesting equipment is not adequate for the collection of logging residues or even full trees. Small tree stems for which the machines are oversized, large hardwood tree tops which are bulky and difficult to transport and to handle, logging residues which are scattered and non-uniform short-rotation plantations, all have their own specific requirements.

Field chipping appears to be consistently more attractive than centralized chipping. Field-chipping equipment is less complex, no foundations or buildings are required and no engineering or design work is called for. There are usually no site acquisition and preparation costs involved (Forestal, 1983).

Figure 70 compares the machine energy requirements of chipping and impact hogging machines producing nominal-sized particles. The energy requirements were adjusted to a common electric motor basis assuming 30% and 80% energy efficiencies for internal combustion and electric engines respectively. It is to be noted that the energy requirement of impact hoggers producing 15 cm size particles is almost equal to that of disc chippers producing 2,5 cm particles. It has been established that biomass should be dried below its fibre saturation point when hammermills and impact hogs are considered as then the energy requirements can be reduced by roughly 50% by such drying (Jones, 1981).

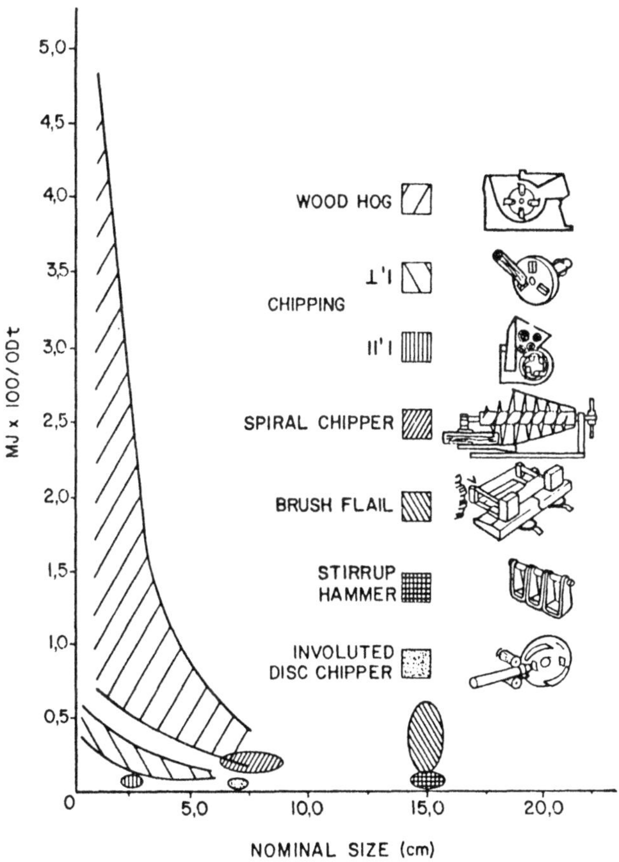

Figure 70. Comparison of machine energy requirements.

Energy plantations have been the subject of intense study since 1974 but interest has waned with the drop in world oil prices. In spite of this, experimental plantations are being established and work is continuing and with the next major increase in oil prices, North America will be better able to meet the challenge of supplying renewable energy.

A very complete study has been conducted by the National Research Council of Canada of existing as well as proposed equipment for the complete mechanization of short-rotation energy plantations. It deals with the site preparation, planting, tending and harvesting of 2-year, 5-year and 10-year-old plantations. It is possible to harvest 18-20 ODt/ha from unfertilized and unirrigated 2-year-old plantations; 42-50 ODt/ha from 5-year-old and 64 ODt/ha from 10-year-old plantations. Different equipment will be required to harvest such plantations at the different rotation ages. For example at 2 years poplar hybrids will weigh about 1,5 kg; at 5 years they will weigh about 25 kg while at 10 they will will weigh about 135 kg (Golob, 1986).

Large-scale wood combustion as an alternative to burning fossil fuels has captured public interest and is receiving continuing support. The burning of wood biomass for energy results in the recycling of CO_2 as compared with the release of stored CO_2 and its release to the environment which occurs when fossil fuels are burned.

12.3 ENERGY PLANTATIONS

Considerable research - in close international collaboration - has been carried on during the last decade to explore and develop energy plantations with forest crops. The studies have included all aspects of such schemes as selection of species and clones, intensive culture with fertilizing and irrigation, harvesting techniques and end use such as pyrolysis for heat production, processing of ethanol and methanol etc. Estimates indicate satisfactory economy at a production level of about 15 t/ha/a which is about half the production of biomass achieved on experimental plots. Large-scale application was started by the Swedish government in 1986. In South Sweden the scheme was subcontracted to a utility firm, which in turn subcontracts plantations to farmers, each planting of at least 10 ha. The estimate of the cost and revenues of this application is shown in Figure 71. As some years will elapse before revenues will accrue, interim financial support to the farmer is provided in the scheme.

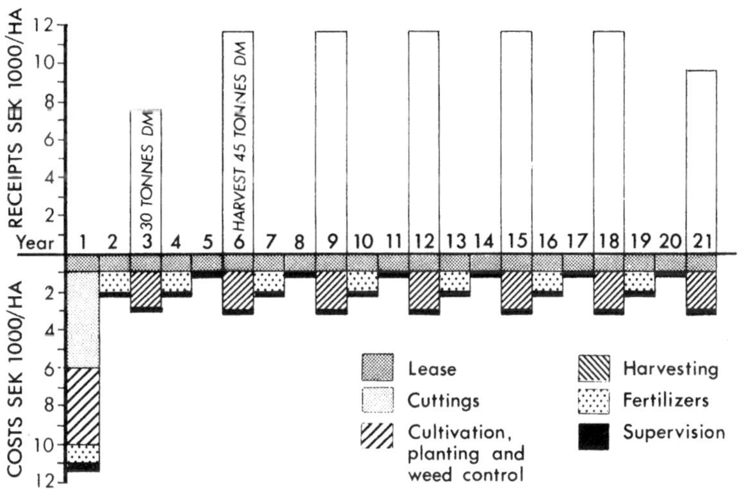

Figure 71. Example of the cash flow for an energy plantation (Sydkraft, 1985)

The Swedish assessment in Figure 71 is based on firm money value in 1985 with a price of 15 SEK/GJ (about 2,50 USD/GJ) for unchipped wood at field edge. The present value of the energy plantation with 21 year rotation exeeds barley cultivation with 7 940 SEK/ha (about 1 300 USD/ha) at 6% rate of interest.

13. FOREST ROADS IN A STAGGERED OWNERSHIP SETTING

When the forest within a natural forest drainage area is split up into many holdings with different owners there is a need for special planning and administrative tools to allow and promote the establishment and continuing maintenance of an adequate forest road network. In some countries, especially those with small forest areas intermixed with agricultural land, the state provides the roads, which then become part of the public road network. Private forest owners may then extend this network, each on his own property. In such cases no special problems arise calling for the attention of forest administrations. However, in many other countries the forest road net is jointly developed, maintained and also financed, at least partly, by the owners of the holdings using or benefitting from the roads. Many times these owners may reach a voluntary agreement on the road matters but generally there are conflicting views on the needs for roads and how costs shall be shared. Then outside expertise is needed to resolve the issue, e.g. the local forest administration.

Table 27. Road network located on forest land in Sweden

	Total length km	Road density, meters/hectare		
		N. Sweden	Mid and S. Sweden	Sweden
Public roads.........	64.600	1.5	3.7	2.2
Private forest roads...	150.750	2.5	10.7	5.1
Total	215.350	4.0	14.4	7.3

As shown in Table 27 the total road system in forest areas in Sweden in the early 1960's was 7,3 m/ha of which only 2,2 or 30% were public roads, the balance being private. These private forest roads are to a large extent road sections or road networks serving many forest holdings. The owners of these holdings within a suitable forest area, together with owners of other real estates, form a road association to administer this very important component for the efficient use of their land assets. A special legislation for private roads is likely to exist in most countries. It may date back to times long before automobiles and trucks came into use, codifying even more ancient customary rules for the provision of roads.

It is common that the state provides subsidies or grants to the road associations, both for road construction cost and for maintenance. This can at least partly be regarded as a compensation for the taxes on fuels and on vehicles which accrue when travelling on private roads but also as support for the promotion of forestry. An obligation to keep the roads open for the public is often tied to such subsidies.

Foresters in many countries are involved in solving road matters, not only for their employer, but also to assist forest owners in solving roading problems on their properties. A presentation of some principles useful in this regard, in addition to the theories of optimal road spacing and standard in Part I, deserves a place in this book.

13.1 LEGISLATION. RIGHT OF WAY

In many countries the public has access to land. People can freely walk or trespass on a forest holding. For example, in the Scandinavian countries there is a customary right for anyone to enter onto any property on foot, bicycle or other non-motorized vehicle, to pick berries, flowers and mushrooms. In other countries the holder of the title of the land can prohibit trespassing. If the title holders of land affected by a road do not unanimously agree to build the road a legal approval is required. In some countries a majority based either on numbers or on area is required whereas in other countries the overall utility and importance of the road are decisive for its legal approval.

Private Road Associations

For the construction and maintenance of a private road with many "shareholders" a road association is often formed. It may be established voluntarily or by a legal procedure with the holdings as members. The task of the association is to be responsible for the construction and continuing maintenance of such road and to handle all the economic and financial matters, including the administration of state subsidies. The association does not as a rule acquire its own equipment but subcontracts all the work to entrepreneurs.

Virgin forests in some countries are often exploited on a short-term lease by a forest enterprise constructing the roads it needs. The land then reverts to the forest administration responsible for the continuing management of the forests. It is often observed that this system is very wasteful. The logger builds the roads he needs during the short time he operates in the area neglecting what comes after. For example, for the drainage of the road he may put in cheap or primitive culverts which disintegrate in a short time. There is often a time lag in the supervision of the roads between the logger and subsequent forest management during which erosion and destruction of the roads may result, e.g. due to malfunction of the installations for drainage. The logger tends to regard his road costs as a variable cost item whereas it should be considered an investment in forest land of considerable magnitude. For example, if logging requires a road density of 15 m/ha for a cost of 20 USD/m it represents an investment in land of 300 USD/ha. The system just described needs to be reformed to ascertain the optimal return on the road investment in the long term.

In Canada and elsewhere in North America the cost of a logging road servicing an area to be cut over is written off as a cost against the wood removed. Such a road is not capitalized. Only if a road is used for an undetermined period is it capitalized. Practice is governed by tax law rather than common sense. Often low-standard, temporary roads are constructed and gradually improved through so-called maintenance to a higher standard - but never capitalized.

The following paragraphs will deal with a situation when a forester is given the task to work out a road plan, to form a road association and to propose how costs shall be shared between the holdings, owned by persons who have conflicting views on the project. First it is necessary to know the requirements for legal approval and other relevant legislation. For example, in the Swedish Act on Private Roads it is stated:

"In a joint road enterprise which is of considerable importance for the rational use of two or several holdings, those holdings shall jointly partake."

"Obligation to partake shall be distributed among the holdings, on which the obligation rests, with regard to the extent of use, to which they are expected to use the road, but greater share may not be placed on any holding than which corresponds to its benefit of the road."

As some of the forest owners are likely to challenge the road plan in order to reduce as much as possible their cost share, it is necessary that the plan conform to some basic principle on which it can be defended. Such principles will be discussed in the following paragraphs.

13.2 CLASSIFICATION OF FORESTS WITH REGARD TO ROAD UTILITY

Forest holdings as a rule show considerable difference in condition and composition bearing on their future road use. They need to be stratified in accordance with such characteristics. The following factors are of main importance: site class, age and stocking of the stands, and the location to existing roads and to the proposed new road. Each holding should be subdivided into area units and each such unit given an index aggregating its properties with regard to these factors. This index shall be an expression of the unit's potential to generate traffic on the road.

A planning period must also be decided. It usually is fixed at 30 to 40 years. The road plan should preferably also contain an assessment of optimal road density and road standards, the principles of which are dealt with in Part I.

Classification into site classes is as a rule an expression of the total increment of wood during a rotation assumed to be directly proportional to expected traffic volume. The age and the stocking of each stratum should be assessed in the

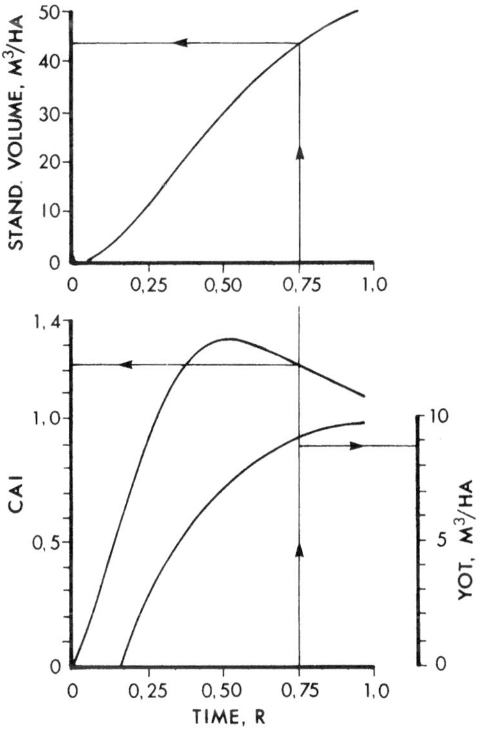

Figure 72. The type scheme in graphical form. It is based on an approximation of stand development in all site classes, reduced to the imaginary site 1 with a mean annual increment (MAI) of 1 m^3/ha during the rotation R. R equals 1,0 and the time scale is thus in ratios of R. Yields of thinning (YOT) are in m^3/ha for thinning intervals of R/10, if longer or shorter to be multiplied by the factor:
actual interval/0,1·R.
Example for site class 8 at 0,75R:

stand volume is	8·45 = 360 m^3/ha
CAI	8·1,22 = 9,8 m^3/ha
YOT	8·8,8 = 70 m^3/ha (if previous thinning was at 0,65·R)

forests but often it is not required to develop too precise figures in this respect. Then the expected traffic generated by each stratum over the planning period must be estimated and discounted to a "present worth" as this figure shall be used for the allocation of the road construction costs on the holdings at the beginning of the planning period. This procedure may seem somewhat complicated but experience has shown that if due consideration to these factors is not taken, the forest owners will be very displeased with the plan as it does not conform with their real future use of the road. However, a method to consolidate these factors will be presented which greatly simplifies the procedure. It is the use of a "type scheme" which describes in a generalized form the development of a forest stand during a rotation, Figure 72. This type scheme is based on work originally suggested by T. Johnsson and A. Modin in a report of a Swedish parliamentary forestry commission (SOU, 1938). It can be very useful for many planning purposes.

The rotation is assumed to terminate when the current annual increment (CAI) goes down to about 80% of its maximum. The thinnings amount to about 40% of the total yield, carried out at optional intervals, normally 2-5 in numbers. The final cut will amount to 50% of the total growth and 10% will be losses from mortality, windthrows etc. The graphs give the quantities in m^3 of wood for a site class with the mean annual increment of 1 m^3/ha/a during the rotation. For other site classes the numbers shall be multiplied with their respective mean annual increment (MAI).

The type scheme is presented here as graphs in Figure 72. With the aid of these graphs the quantities harvested during the planning period can be derived and discounted to an aggregated present worth, which is an expression of the traffic volume (TI) that the stand generates. The sum of these TI's for the stands of a forest holding gives a number to its use of the road.

In Table 28 a middle-aged stand on average site class and stand density 0,7-1,0 is given the relative value of 1,0 for the traffic volume it generates over a planning period of 30-40 years, discounted to the year 0 (= the year of road construction). The traffic volume of other stands is expressed as ratios. For a more detailed description of this procedure (Sundberg, 1965).

Table 28. Traffic indices (TI) for the grading of the present worth traffic volume per unit area of a forest stand.

Stand Description	Bare land, juvenile forests			Young forests			Middle-aged forests			Mature forests		
Density	< 0.7	0.7—1.0	> 1.0	< 0.7	0.7—1.0	> 1.0	< 0.7	0.7—1.0	> 1.0	< 0.7	0.7—1.0	> 1.0
Site Class			For each higher site class multiply by 1.33									
Better	0.5	0.5	0.7	0.8	0.9	1.1	1.1	1.3	1.6	1.3	1.7	2.1
Average	0.4	0.4	0.5	0.6	0.7	0.8	0.8	1.0	1.2	1.0	1.3	1.6
Poorer	0.3	0.3	0.4	0.4	0.5	0.6	0.6	0.8	0.9	0.8	1.0	1.2
			For each poorer site class multiply by 0.75									

Explanations to table 28

Stand: Described in four classes, distinguished as follows.

Bare land, juvenile forest: no stand or stand with average height below breast height.

Young forests: stand with average height above breast height, but for which the current annual growth has not yet culminated (upper age in Sweden, north 60—70 years, middle 50 years, south 30—40 years).

Middle aged forests: stands from the age of the culmination of current annual growth to the age of economical maturity.

Mature forests: stands above the age of economical maturity.

Density: 1.0 = optimum economical stocking, 0.7 = 70 % of opt. ec. stocking.

Site Class: "Average" = the dominant site class within the drainage area of the road. "Better" and "Poorer" are based on the *Jonsson* site grading system, for which the difference between two classes is 25 % in average annual growth during a rotation. (Per cent difference is based on the site class with the higher production.)

13.3 LOCATION

In Part I (Chapter 2.7) two functions of a forest road were distinguished: crosswise for the extraction of timber and lengthwise for its transport on the road. The location of each stand shall be enumerated in these two respects. It is obvious that a stand close to the road enjoys a greater benefit from it than does a stand located more distant from it. To simplify the identification of the crosswise location the area can be subdivided in equidistant zones and an estimate be made of the relative crosswise benefit for each zone. An example is shown in Table 29 with ratios of the benefit of the nearest zone with benefit 1,0.

The lengthwise function is identified as the distance from the landing(s) to the juncture of the road with a (public) road, L_i.

If wood assortments are hauled in different directions each fraction shall be enumerated separately.

Table 29. Correction factors F_z for the grading of the crosswise function of a road with regard to the crosswise location of the stand (unit area)

Distance-zone	Correction factor, F_z
< S	1,0
S - 2S	0,6
2S - 3S	0,4
3S - 5S	0,3
5S	0,2

Example: Assume a young forest stand of 2 ha on a poor site class with a density of 0,7-1,0 located in distance zone S-2S from the road and 3 km from its junction with a public road. From Table 28 we get TI = 0,5 and from Table 29 F_z=0,6. The nominators in the formulae are for this stand:
for road cost to minimum standard $2 \cdot 0,5 \cdot 0,6 = 0,6$
for standard improvement cost $2 \cdot 0,5 \cdot 3 = 3$

13.4 COST DISTRIBUTION

It is evident that the crosswise function is fully met by a road of minimum standard as further explained in Part I. The lengthwise function is improved by road standards above this minimum. Consequently, it is logical to allocate the construction cost of the road to minimum standard in proportion to the crosswise benefit enjoyed by each stand, disregarding that section of the road which is built to a higher standard. The costs of standard improvements are then allocated according to lengthwise use. The formulae are:

Share of stand "i" (with the area A) of total cost to minimum standard =

$$= \frac{A_i \cdot TI_i \cdot F_{z,i}}{\Sigma A_i \cdot TI_i \cdot F_{z,i}}$$

and for the cost of standard improvements =

$$= \frac{A_i \cdot TI_i \cdot L_i}{\Sigma A_i \cdot TI_i \cdot L_i}$$

These shares are then aggregated for each holding and the sum is its share of the respective two cost items.

13.5 ROAD MAINTENANCE COST

The cost distribution of road construction is as a rule also applicable for road maintenance. A review could be made after some years to adjust to eventual changes. Maintenance cost can also be covered by road tolls but as these are both difficult to control and cumbersome to administer, tolls are seldom used except for external, occasional users.

13.6 CONCLUDING REMARKS

There are other possibilities for the distribution of road cost than the approach suggested in the previous paragraphs. The "Tonnekilometer" method distributes costs according to the transport work of the users on the road. It will result in a high cost share on users in the more remote parts of the road system. As it disregards the crosswise function of the road, a modification is often required to reduce the impact of distance on the result. Another approach is to assess the operating cost of each holding before and after the road construction and allocate the cost in proportion to the gains for each holding. No method can be given preference on theoretical grounds. The main point is that the proposal agrees with what the members consider as just. If there are many holdings in a road association the calculations and the continuing administration can be computerized (Sundberg & Turén, 1971). Some years ago an association was established in a sparsely populated area in Sweden encompassing about 118 km of roads with 1249 holdings as members.

REFERENCES

Addor, E. E. (1963). The effect of stem spacing on vehicle performance. Military evaluation of geographic areas. Reports on activities to April 1963 - Dec. 1963. U.S. Army Engineer Waterways Experiment Station, Vicksburg, Mississippi.

Ager, B. (1972). Träd- och kvistdata för skogstekniska kalkyler. Skogshögskolan. Rapporter och Uppsatser, Nr. 50. Sweden.

Andersson, S., and Björheden, R. (1986). Skogsbränsleuttag/ Ny teknik (Production of Wood Fuel/New Techniques, in Swedish only). Statens energiverk, Stockholm.

Anon (1929). Public Roads, Volume 10, p. 181. December.

Anon (1961). Forest work and earth mover tires. Forestry Equipment Note B.27.61. Food and Agricultural Organization, Rome. February.

Anon (1965). Research plan for the development of a quantitative cross country mobility prediction system. U.S. Army Engineer Waterways Experiment Station, Vicksburg, Mississippi. June.

Anon (1967). Dimensions des Grume et choix d'un equipment. Etude de repartition des grumes par dimension. A6 bis-Centre Technique Forestier Tropical. Juillet.

Anon (1976). Rome Shear Head application on 80 to 100 HP wheel loaders. Rome job report, Rome Industries. Cedartown, Georgia.

Anon (1982). Log loading methods and costs in the Tennessee Valley Region. TN-B46. TVA/ONR/LFR 82/7. April.

Armson, K.A. (1977). Forest Soils. University of Toronto Press.

Axelsson, S. A. (1972). Logging Machine Failure Avoidance: Identification and Measurement of Key Parameters. Logging Research Report No. 48. Pulp and Paper Research Institute of Canada, Pointe Claire, Quebec, Canada. October.

Bekker, M.G. (1950). Relationship between soil and a vehicle. S.A.E. Reprint January 9.

Bekker M.G. (1961). Evaluation and Selection of Optimum Vehicle Types under Ramdon Terrain Conditions. General Motors Defense Systems Division, Land Mobility Laboratory, No. ER-61-107. April.

Bendz, M. and Järvholm, A. (1970). Logging and Transport in Tropical High Forest. Research Note No. 38. Royal College of Forestry, Stockholm, Sweden.

Bennett, W.D., Winer, H.I., and Bartholomew, A. (1965). Measurement of Environmental Factors and Their Effect on the Productivity of Tree Length Logging with Rubber Tired Skidders. Woodlands Section Index No. 166, Pulp and Paper Research Institute of Canada, Pointe Claire, Quebec.

Bjerkelund, T.C. (1971), Off-Road Transportation Systems Currently used for Steep Terrain Conditions in Eastern Canada. Pulp and Paper Magazine of Canada, Volume 72, No. 12. December.

Bjerkelund, T.C. (1983). Circular Saw Tree Felling Devices: A Promising New Development in Eastern Canada. Meeting Division 3, IUFRO, Jönköping, Sweden. June.

Blackmon, C.A. and Randolph, D.D. (1968). An Analytical Model for Predicting Cross-Country Vehicle Performance. Appendix B: Vehicle Performance in Lateral and Longitudinal Obstacles (Vegetation) - Longitudinal Obstacles. Technical Report No. 3 - 783. U.S. Army Engineers Waterways Experiment Station, Vicksburg, Mississippi. December.

Blackmon, C.A. and Stoll, J.K. (1968). An Analytical Model for Predicting Cross-Country Vehicle Performance. Appendix B: Vehicle Performance in Lateral and Longitudinal Obstacles (Vegetation) - Lateral Obstacles. Technical Report No. 3 - 783. U.S. Army Engineers Waterways Experiment Station, Vicksburg, Mississippi. December.

Boyd, J.H. and Novak, W.P. (1977). A Method of Comparing Logging System and Machine Concepts. Proceedings Woodlands Section Canadian Pulp and Paper Association Annual Meeting, Montreal, Canada. March.

Boyd, J.H. (1977). A Status Report: New Machines and Systems Research and Development. Forest Engineering Research Institute of Canada. Pointe Claire, Quebec, Canada. March.

Bredberg, C.J. and Wästerlund, I. (1983). Wurzel- und Bodenschäden durch Fahrzeuge. Forstwissenschaftliches Zentralblatt 102 Jahrgang (1983), H.2, S86-98.

Bredberg, C.J. (1984). Technical Developments for Improving the Silvicultural Quality of Thinnings - mimeo. Department of Operational Efficiency, College of Forestry, Swedish University of Agricultural Sciences. Garpenberg.

Brown, H.P., Panshin, A.J., Forsaith, C.C. (1952). Textbook Wood Technology Vol. 2. McGraw-Hill Book Company, New York.

Bruun, E. (1986). Mini Bruunett brochure, Bruun System Ab, Filipstad, Sweden.

Caterpillar, (1976). Caterpillar Performance Handbook. Edition 6, Peoria, Illinois.

Chardin, A., Le Pitre, C., LeRay, J. (1959). Dimensions des Grumes et Choix d'un Equipment. Revue Bois et Forêts des Tropiques, Number 65. Mai-Juin.

Chatarpaul, L., Burgess, D. and Hendrickson, O.Q. (1984). A Comparison of Biomass and Nutrients Removed in Whole Tree Harvesting and Conventional (Stem only) Harvesting. In Hasnain, S., ed. Proc. Fifth Cdn. Bioenergy R & D Seminar, Elsevier, London.

Chinn, G.P. (1982). Road Transport of Tree Lengths and Full Trees in Canada. IUFRO Div. 3 Congress, Munich.

Clark, A. III and Taras, M.A. (1976). Biomass of Shortleaf Pine in a Natural Sawtimber Stand in Northern Mississippi. U.S.D.A. Forest Service Research Paper SE-146, Southeastern Forest Experiment Station, Asheville, North Carolina. April.

Clark, A. III and Schroeder, J.G. (1977). Biomass of Yellow-Popar in Natural Stands in Western North Carolina. U.S.D.A. Forest Service Research Paper SE-165, Southeastern Forest Experiment Station, Asheville, North Carolina. September.

Cline, Ch. (1967). Wood Tick Trail. Pulpwood Production. December.

Conway, S. (1977). How to cope with smaller logs and their higher harvesting costs. Forest Industries. September.

Cubbage, F.W. and Werblow, D.A. (1985). Trends in Forest Harvesting Equipment Costs. ASAE Paper No. 85-1591. Chicago, Illinois. December.

Den Uyl, D. (1968). Forest Decision Making in an Integrated Firm. Journal of Forestry. October.

Drolet, J.C., Newnham, R.M., and Tsay, T.B. (1971). Branchiness of Jackpine, Black Spruce and Balsam Fir in Relation to Mechanized Delimbing. Forest Management Institute , Canadian Forestry Service, Information Report FMR-X-34. June.

Elpatievsky, M.P., Rumiantsev, S.P., Yormolovich, B.K. (1934). Methods of Slash Disposal Adopted to Forest Types. Translation 281, Division of Silvics, U.S.D.A. Forest Service, Washington, D.C.

Ergon, N.D. (1981). Relifing of Heavy Equipment. Proceedings 62nd Annual Meeting Woodlands Section, Canadian Pulp and Paper Association, Toronto. March.

FAO (1985). World Forest Resources, 1980.

Fellows, E.S. (1974). Report for Canadian International Development Agency on Peru Forest Sectoral Study. Project 730-00142. Ottawa. November.

Folkema, M.P. (1984). Circular Saw and Cone Saw Felling Heads: An Update. Technical Report No. TR-56, Forest Engineering Research Institute of Canada. Pointe Claire, Quebec. March.

Forestal International Limited (1983). Guidelines for assessing profitability of chipping forest biomass at field and central locations. Energy Mines and Resources Canada, Ottawa.

FPRS (1982). Harvesting Small Timber: Waste Not, Want Not. Proceedings 7314, Forest Products Research Society, Madison, Wisc., USA.

FPRS (1983). The Small tree resource: A materials handling challenge. Proceedings 7306, Forest Products Research Society, Madison, Wisc., USA.

FPRS (1984). Harvesting the South's Small Trees. Proceedings 7340, Forest Products Research Society, Madison, Wisc., USA.

Froelich, H.A., Azevedo, J.,Cofferata, P., Lysne, D. (1980). Predicting Soil Compaction on Forested Land. U.S.D.A. Forest Service Equipment Development Centre. Missoula, Montana. December.

Gabriel, W.J., Nissen, R.L., Burne, K. (1974). A Time Study of Top-Logging Northern Hardwoods in New York State. Applied Forestry Research Institute Report No. 23, College of Environmental Science and Forestry, State University of New York, Syracuse, N.Y.

Garlicki, A.M., and Calvert, W.W. (1969). A Comparison of Power Requirements for Full-Tree Versus Tree-Length Skidding. Pulp and Paper Magazine of Canada. July.

Gauther, S.J. (1968). Cost Control in Woodlands Operations, Woodlands Section Index No. 2475 (6-5) Canadian Pulp and Paper Association, Montreal, Canada, September.

Giedion, S. (1948). Mechanization Takes Command. W.W. Norton and Company, Inc.

Golob, T.B. (1986). Analysis of short rotation forest operations. Division of Energy, National Research Council of Canada. Publ. NRCC No. 26014. Ottawa.

Gonsior, M. (1986). Mechanized Harvsting and Processing in Mountainous Terrain of Western Montana: A Case Study. Ninth Annual Meeting of The Council on Forest Engineering, Mobile, Alabama. September.

Goodyear (1970). Off-the-road Truck Tires. Engineering Data. The Goodyear Tire and Rubber Company, Akron, Ohio. 44316.

Hakkila, P. (1971). Coniferous Branches as a Raw Material Source. Communications Instituti Forestalis Fenniae 75, Helsinki, Finland.

Hakkila, P. (1988). Harvesting and Utilization of Forest Biomass. Springer-Verlag Berlin, Heidelberg, New York.

Hakkila, P., Leikola, M. and Salakari, M. (1979). Production, Harvesting and Utilization of Small-sized Wood. The Finnish National Fund for Research and Development. Sarja B No 46b, Helsinki.

Hallett, R.M. and McCraw, W.E. (1970). Studies on The Productivity of Skidding Tractors. Departmenty of Fisheries and Forestry. Canadian Forestry Service. Publication No. 1282, Ottawa, Canada.

Hallonborg, U. (1979). Spårdjup efter skotare. Forskningsstiftelsen Skogsarbeten. Stockholm. Teknik Nr. 4.

Hanley, D.P. (1976). Tree Biomass and Productivity Estimated for Three Habitat Types of Northern Idaho. Bulletin No. 14, College of Forestry Wildlife and Range Sciences, University of Idaho, Moscow, Idaho. May.

Harkness, W.D. (1963). Control in Logging. Seminar on Mechanized Logging, Luleå, Sweden. August.

Hazzard, H. (1969). A Modern Two Cubic Inch 6-5 lb. Chainsaw. ASAE Paper 690599. Milwaukee. September.

Hedbring, O., Nilsson, P.O., Åkesson, H. (1966). Analys av några avverkningssystem för gallring. Skogsarbeten, Stockholm.

Herschelman, J.W. and Domenech. D.W. (1980). Recovery of Above-Ground Woody Biomass Using Operational Modifications of Conventional Harvesting Systems. Report DOE/ET/20045-T1. Department of Energy, Washington, D.C.

Hertig, B.A. (1979). Ergonomics. Fundamentals of Industrial Hygiene, National Safety Council, Chicago, Illinois.

Hogentogler, C.S. (1937). Engineering Properties of Soils. McGraw-Hill Book Company.

Håkansson, S.-G. (1977) Space Requirements for Forwarders in Thinning. Report Nr. 114, Department of Operational Efficiency, College of Forestry, Swedish University of Agricultural Sciences. Garpenberg, Sweden.

International Business Machines (1961). Introduction to Control Systems. General Information Manual.

Janosi, Z.J., Liston. R.A., Martin, L.A., Sloss, D.A. (1970). Commercial Off-Road Vehicles. S.A.E. Paper 700012, S.A.E. Automotive Engineering Congress. Detroit, Michigan. January.

Jiles. R.A. and Lehman, J.W. (1960). Hardwood Logging Methods and Costs in the Tennessee Valley. TVA Report No. 232-60. Norris. Tennessee. April.

Johnston, J.S. (1967). Investigations of Some Variables in the Crosscutting of Small Logs by Shear Blades. Proceedings IUFRO, Munich, Germany. September.

Johnston, J.S. (1972). Effect of Superimposed Vibrations on a Shear Blade While Crosscutting Wood. Internal Report OP-51 EFPL. Canadian Forestry Service. January.

Johnston, J.S. and St.Laurent, A. (1974). Evaluation of Damage in Shear Felled Sawtimber. EFPL, Ottawa, Canada. August.

Johnston, J.S. and St.Laurent, A. (1975). Superimposed Vibrations on a Tree Shear Blade. Report OPX-138E. Eastern Forest Products Laboratory, Ottawa, Canada.

Jones, K.C. and Associates (1981). Field tests to develop energy saving wood comminution techniques. Forest Engineering Research Institute of Canada, Special Report No. SR-15, Pointe Claire, P.Q.

Jullander, I. and Stockman, L. (1978). The Life Cycle of Wood. Information No. 110-1978. National Swedish Board for Technical Development. Stockholm.

Keays, J.L. (1976). Forest Harvesting of the Future. AICLE Symposium Series 72 (157) 4-12.

Keen, R. (1963). Weight and Centres of Gravity Involved in Handling Pulpwood Trees. Pulp and Paper Research Institute of Canada, Technical Report No. 340. Pointe Claire, Quebec. November.

Kempe, C. (1964). The Forces and Damage to Fibre Structure Involved in Hydraulic Shearing of Round Wood. Thesis for Licentiate Degree, Royal Swedish College of Technology, Stockholm.

Koning, J.W., and Skog, K.E. (1987). Use of wood energy in the United States - an opportunity. Biomass 12, 27-36.

Koplow, R.A. (1974). What Needs to be Done to Achieve a Healthy Pulpwood Supply. Paper Trade Journal. May 6.

Koroleff, A.M. (1954). Full-Tree Logging - A Challenge to Reserarch. Pulp and Paper Research Institute of Canada. Pointe Claire, Quebec.

Koten, D.E. (1969). Engineered Harvesting Systems - Problems and Promises. Meeting of New York Section, Society of American Foresters, Liberty, New York.

Kurelek, J. (1976). Economics and Productivity of Multi-function Forest Harvesting Machines. Proceedings 57th Annual Meeting, Woodlands Section, Canadian Pulp and Paper Association, Montreal.

Kurelek, J. and Denovan, J.J. (1969). Koehring Harvester For Pulpwood. ASAE Paper 690554. August.

Lampen, S., Polkki, V., Salminen, J. (1970). Kockum Bruunett KL-836B Forwarder. Metsätehon Katsaus, Metsäteho, Helsinki. November.

Lanly, J.P. (1982). Tropical Forest Resources. FAO Forestry Paper 30, Rome.

Legendre, C. (1973). Improving Productivity = Expensive Hardware, Better Qualified Workers, How About the Organization? Woodlands Section, Canadian Pulp and Paper Association Proceedings 54th Annual Meeting, Montreal. March.

Lenhard, R.J. (1986). Changes in Void Distribution and Volume During Compaction of a Forest Soil - Soil Science of America Journal, 50 (2) 462-464.

Li, C.Y. (1956). Basic Concepts on the Compaction of Soil. Proceedings American Society of Civil Engineers. Soil Mechanics and Foundation Division. Paper 862.

Löffler, H.J. (1984). Terrain Classification for Forestry. FAO/ECE/ILO Joint Committee on Forest Working Techniques and Training of Forest Workers. TIM/EFC/WP.1/R.51. August.

Lull, H.W. (1959). Soil Compaction on Forest and Range Lands. U.S.D.A. Forest Service Misc. Publication No. 768, Washington, D.C. January.

Lussier, L.J. (1961). Planning and Control of Logging Operations. Université Laval, Quebec, Canada.

Lutz, J.F. (1952) Mechanical Impedance and Plant Growth. In Soil Physical Conditions and Plant Growth. Academic Press, New York, N.Y.

McColl, B.J. and Pepler, W.A.E. (1953). The Development of Mechanical Pulpwood Logging Methods for Eastern Canada. Woodlands Section, Canadian Pulp and Paper Association, W.S. Index No. 1325. July.

McCraw, W.G. and Hallett, R.M. (1970). Studies on the Productivity of Skidding Tractors. Eastern Forest Products Laboratory, Canadian Forestry Service, Ottawa, Publication No. 1282.

McCraw, W.G. and Silversides, C.R. (1970). Analysis of Tree Harvesting Machines and Systems - A Methodology. Forest Management Institute, Canadian Forestry Service. Information Report FMR-X-27. July.

McLauchlan, T.A. and Kusec, D.J. (1975). Ribbed Blades Reduce Tree Shearing Damage. Canadian Forest Industries. February.

Maleta, B.P. and Sood, K.G. (1984). Physiological Strains of Forest Worker (A Case Study in Tropical Hardwood Forests of Central India). Indian Forester (1984) 110 (4) 337-344.

Martin, R.B. (1965). Studies on Articulated Rubber-Tired Vehicles. International Society for Terrain Vehicle Systems US-Canadian Regional Conference. August.

Mellgren, P.G. (1978). Optimal Logging Machine Systems. Pulp and Paper Canada. February.

Mellgren, P.G. (1987). Selecting the Appropriate Wood Harvesting Machines. Proceedings Woodlands Section, Canadian Pulp and Paper Association Annual Meeting, Montreal. March.

Miller, W. (1978). Crosscut Saw Manual. USDA Forest Service. Missoula Equipment Development Center. Missoula, Montana. June.

Mills, H. L. (1962). Quantitative Environmental Studies, South Florida in Military Evaluation of Geopgraphic Areas, Reports on Activities to April 1963. U.S. Army Corps of Engineers Waterways Experiment Station, Vicksburg, Mississippi.

Nataraj, A.R. (1968). Terrain Trafficability Evaluation Using Aerial Photographs: A modified approach. International Institute For Aerial Survey and Earth Sciences (ITC), Delft, Netherlands. July.

Nemtsov, V.P. (1975). Influence of Natural Conditions on Selection Parameters of Felling Machines. FAO/ECE/ILO Symposium on Multipurpose Logging Machines. Sweden. June 10-18. Tim/EFC/WP.1/SEM.1/R.23.

Newnham, R.M. (1968). Simulation Models in Forest Management and Harvesting. Forestry Chronicle. February.

Newnham, R.M. (1971). Simulation Studies in the Logging Development Program of the Canadian Forestry Service - A Progress Report. Forest Management Institute, Canadian Forestry Service, Ottawa. April.

Newnham, R.M. and Sjunnesson, S. (1969). A Fortran Program to Simulate Harvesting Machines for Mechanized Thinning. Forest Management Institute, Canadian Forestry Service, Information Report FMR-X-25.

Nordfjell, T. (1988). Minilunnare - användare och driftserfarenheter 1987. Dept of Operational Efficiency. Stencil Nr 96. Swedish University of Agricultural Sciences, Garpenberg.

Nuttall, C.J. Jr. (1967). Ground Crawling: 1966. The State-of-the-Art of Designing Off-Road Vehicles. Contract Report No. 3-162, U.S. Army Corps of Engineers Waterways Experiment Station, Vicksburg, Mississippi.

Olsen, E.D. and Gibbons, D.J. (1983). Predicting Skidder Productivity: A Mobility Model. Research Bulletin 43, Forest Research Laboratory, School of Forestry, Oregon State University, Corvallis, Oregon, August.

Osmaston, F.C. (1968). The Management of Forests. George Allen & Unwin Ltd., London, England.

Ouellet, D. (1983). Biomass Prediction Equations for Twelve Commercial Species in Quebec. Information Report LAU-X-62E. Laurentien Forest Research Centre, Canadian Forestry Service, Ste-Foy, Quebec.

Palm, I., Wessgren, H, Wiklund, M. (1968) Bearbetningsoperationer vid mekaniserad skogsavverkning. Stockholm. April.

Pickering, R.R. (1975). Logma Limbs All. Pulp and Paper Canada, Volume 76, No. 9. September.

Piette, G. (1952). Bearing Capacity of Cohesive Soils. Roads and Engineering Construction. November.

Radforth, N.W. (1952). A Suggested Classification of Muskeg For The Engineers. Technical Memorandum No. 24, National Research Council, Ottawa, May.

Ralston, K. (1970). Skidder Tire Selection Guide. American Pulpwood Association Technical Release 70-R-7.

Rouse, J.J. (1968). Woodlands Cost Control - Is Responsibility Accounting the Answer? Canadian Pulp and Paper Association Woodlands Section Index No. 2477 (B-5). April 19.

Rollerson, T.P. and Kennedy, C.A. (1984). Compactive Effects of a Low Ground Pressure Skidder. Woodland Services Division, MacMillan-Bloedel Ltd., Vancouver, B.C. August.

Rybczynski, W. (1985). Taming The Tiger - The Struggle to Control Technology. Viking Penguin Inc., New York, N.Y.

St.Laurent, A. (1975). Memorandum Trip to Price Brothers, Lac St. Jean Area. Eastern Forest Products Laboratory, Canadian Forestry Service, Ottawa Internal Report. February.

Sarles, R.L. and Luppold. W.G. (1986). Technoeconomic Analysis of Conventional Logging Systems Operating From Stump to Landing - Research Paper NE-577, U.S.D.A. Forest Service, Northeastern Forest Experiment Station, Princeton, West Virginia.

Sayn-Wittgenstein, L. and Nielsen, U. (1972). Development of Forest Technologies in the Member Countries of the Andean Pact. Forest Management Institute, Canadian Forestry Service, Ottawa. June.

Scherman, S. (1987). How is a New Machine System Introduced and Evaluated? In Proceedings of Harvesting Machines and Systems Evaluation Workshop. Dept. of Operational Efficiency, Res. Notes 108, Swedish University of Agricultural Sciences, Garpenberg.

Silversides, C.R. (1960). The Influence of Tree Characteristics on Logging Efficiency and Cost. Pulp and Paper Magazine of Canada. August.

Silversides, C.R. (1967). Operations Performed in Forest Stands. Proceedings of Symposium on the Economic Location of Working Operations. FAO/ECE/ILO Joint Committee on Forest Working Techniques and Training of Forest Workers. Geneva, April.

Silversides, C.R. (1968). The Economics and Techniques of Timber Harvesting. Ninth Commonwealth Forestry Conference, India.

Sirois, D.L. (1986). A Mobile Harvester for Utilization of Weed Trees and Residues. Proceedings of Conference. Weed Control in Forest Management. Purdue University. West Lafayette, Indiana.

SLU (1984). Forest energy in Sweden. Dept of Operational Efficiency, Swedish University of Agricultural Sciences, Garpenberg.

SLU (1987). Skogen i energiförsörjningen (The forest for energy supply. In Swedish only). Skogsfakta Nr 10. Swedish University of Agricultural Sciences, Uppsala.

Smith, L.A. (1980). Assessment of the Ergonomics Literature to Identify Productivity Improvement, Applications and Research Needs Partinent to Southern Forest Harvesting. Department of Industrial Engineering, Auburn University, Alabama, 36849. October.

SOU (1938). Improving Rational Wood Production. (In Swedish only). Statens Off. Utredningar. SOU 1938:58, Stockholm.

Spangler, M.G. (1951). Engineering Aspects of Soils. Society of Automotive Engineers. April.

Spanjer, D.L. (1968). Consistency in the Measurement of Logging Production and Cost. Forest Engineering Conference, ASAE, Michigan State University, East Lansing, Michigan. September.

Sundberg, U. (1954). The Costs of Cutting and Transport of Small Sized Wood. (In Swedish only). In Klenvirke, SOU 1954: 29, p 24-39. Stockholm.

Sundberg, U. (1956). The Use of Power Saws in Forest Operations. FAO/ECE/ILO Joint Committee on Forest Working Techniques and Training of Forest Workers. Geneva.

Sundberg, U. (1959). A Study on the Manual Handling of Wood. Medd. Statens Skogsforskningsinstitut, Bd 49, Nr 2. Stockholm.

Sundberg, U. (1965). Distribution of the Costs of Joint Forest Roads According to Crosswise and Lengthwise Road Functions. Studia Forestalia Suecica Nr 30.

Sundberg, U. and Turén, A. (1971). The TL-Method for Distribution of the Costs of Joint Forest Roads, with a Computer Programme. (In Swedish with English Summary). Report No. 47. Dept. of Operational Efficiency, Swedish University of Agricultural Sciences, Garpenberg.

Sydkraft (1985). Energy Forestry - the new farm crop. Malmö.

Taras, M. A. and Clark A. III (1974). Above-Ground Biomass of Loblolly Pine in a Natural Uneven-aged Sawtimber Stand in Central Alabama. Proceedings TAPPI Forest Biology Conference.

Terzaghi, K. (1944). Theoretical Soil Mechanics. J. Wiley, New York, N.Y.

Thomas, J.B. (1978). Biomass Inventory of Tolerant Hardwoods. ENFOR P-21, Canadian Forestry Service, Ottawa, Canada. December.

Thomson, J.G. (1968). Off-Road Transportation. Faculty of Forestry, University of New Brunswick, Fredericton, N.B.

Tufts, R.A., Izlar, R., Simmons, D. and Thurber, J. (1981). Forestry Equipment Cost Index: 1968-1978. Southern Journal of Forestry 5 (4) : 201-204.

Urwick, L. (1947). The Elements of Administration. Sir Isaac Pitman and Son Ltd., London.

U.S.D.A. (1940). Wood Handbook. Forest Products Laboratory. Washington, D.C. June.

Wackerman, A.E., Hagenstein, W.D. and Michell, A. (1966). Harvesting Timber Crops. McGraw-Hill Book Company, New York. American Forestry Series.

Wakefield, W.E. (1932). Investigation Into The Various Factors Governing The Efficiency of Bucksaws. Proceedings Annual Meeting, Woodlands Section, Canadian Pulp and Paper Association, Montreal. February.

Westoby, J.C. (1966). Possibilities for the Developing Countries. UNESCO Committee on Application of Science and Technology to Development. Rome. October.

Widule, C.J., Foley V., and Demo, G. (1978). Dynamics of the Axe Swing. Ergonomics, Volume 21, Number 11, 925-930.

Wiklund, M. (1967). Forces and Damage in Felling and Bucking with Knives and Shear Type Tools. Report No. 9. Skogsarbeten, Stockholm.

Wilson, B.F. (1970). The Growing Tree. University of Massachusetts Press, Amherst, MA.

Whittaker, R.H., and Woodwell, G.M. (1971). Productivity of forest ecosystem. pp 169-175. UNESCO, Paris.

Young, H.E. (1974). Biomass, Nutrient Elements, Harvesting and Chipping in the Complete Tree Concept. Journal of the Association of Consulting Foresters, Volume 19, No. 4. October.

Zivnuska, J. (1966). The Integration of Forest Development Plans and National Development Plans. How to Make the Forestry Case at the National Level. Sixth World Forestry Congress, Madrid, Spain. June.

MIX
Papier aus verantwortungsvollen Quellen
Paper from responsible sources
FSC® C105338

If you have any concerns about our products,
you can contact us on
ProductSafety@springernature.com

In case Publisher is established outside the EU,
the EU authorized representative is:
Springer Nature Customer Service Center GmbH
Europaplatz 3, 69115 Heidelberg, Germany

Printed by Libri Plureos GmbH
in Hamburg, Germany